MOTORS
for MAKERS

A Guide to Steppers, Servos, and Other Electrical Machines

Matthew Scarpino

que®

800 East 96th Street
Indianapolis, Indiana 46240

MOTORS FOR MAKERS: A GUIDE TO STEPPERS, SERVOS, AND OTHER ELECTRICAL MACHINES

ISBN-13: 978-0-13-403283-2

ISBN-10: 0-13-403283-7

Library of Congress Control Number: 2015951266

Printed in the United States of America

First Printing: December 2015

Trademarks

All terms mentioned in this book that are known to be trademarks or service marks have been appropriately capitalized. Que Publishing cannot attest to the accuracy of this information. Use of a term in this book should not be regarded as affecting the validity of any trademark or service mark.

Warning and Disclaimer

Every effort has been made to make this book as complete and as accurate as possible, but no warranty or fitness is implied. The information provided is on an "as is" basis. The author and the publisher shall have neither liability nor responsibility to any person or entity with respect to any loss or damages arising from the information contained in this book.

Special Sales

For information about buying this title in bulk quantities, or for special sales opportunities (which may include electronic versions; custom cover designs; and content particular to your business, training goals, marketing focus, or branding interests), please contact our corporate sales department at corpsales@pearsoned.com or (800) 382-3419.

For government sales inquiries, please contact governmentsales@pearsoned.com.

For questions about sales outside the U.S., please contact international@pearsoned.com.

Editor-in-Chief
Greg Wiegand

Executive Editors
Rick Kughen
Bernard Goodwin

Development Editor
Greg Kettell

Managing Editor
Kristy Hart

Senior Project Editor
Betsy Gratner

Copy Editor
Bart Reed

Indexer
Lisa Stumpf

Proofreader
Leslie Joseph

Reviewers
John Baichtal
Bryan Bergeron
Rich Blum
Stephen Hobley
James Floyd Kelly
Pete Prodoehl
Paul Tan

Publishing Coordinators
Michelle Housely
Cindy Teeters
Kristen Watterson

Cover Designer
Mark Shirar

Compositor
Nonie Ratcliff

CONTENTS AT A GLANCE

CONTENTS

ABOUT THE AUTHOR

Matthew Scarpino is an engineer with more than 12 years of experience designing hardware and software. He has a master's degree in electrical engineering and is an Advanced Certified Interconnect Designer (CID+). He is the author of *Designing Circuit Boards with EAGLE: Make High-Quality PCBS at Low Cost*.

WE WANT TO HEAR FROM YOU!

As the reader of this book, *you* are our most important critic and commentator. We value your opinion and want to know what we're doing right, what we could do better, what areas you'd like to see us publish in, and any other words of wisdom you're willing to pass our way.

We welcome your comments. You can email or write to let us know what you did or didn't like about this book—as well as what we can do to make our books better.

Please note that we cannot help you with technical problems related to the topic of this book.

When you write, please be sure to include this book's title and author as well as your name and email address. We will carefully review your comments and share them with the author and editors who worked on the book.

Email: feedback@quepublishing.com

Mail: Que Publishing
            ATTN: Reader Feedback
            800 East 96th Street
            Indianapolis, IN 46240 USA

READER SERVICES

Register your copy of *Motors for Makers* at informit.com for convenient access to downloads, updates, and corrections as they become available. To start the registration process, go to informit.com/register and log in or create an account.* Enter the product ISBN, 9780134032832, and click Submit. Once the process is complete, you will find any available bonus content under "Registered Products."

*Be sure to check the box that you would like to hear from us in order to receive exclusive discounts on future editions of this product.

INTRODUCTION

When I received my master's degree in electrical engineering in 2002, I couldn't help but feel a little disappointed. I knew all about analog circuit theory, but I knew next to nothing about practical circuit boards. I could compute the Lorentz force in an electric motor, but I had no idea how motor controllers worked in the real world. Put simply, I could write programs and solve equations, but I couldn't *make* anything.

Shortly after I received my degree, the first Arduino boards appeared in the marketplace. Their simplicity and low cost sparked a worldwide interest in electronics, and within a few years, the Maker Movement was born. Makers aren't interested in heavy mathematics and physics. Makers are concerned with what they can build. Whether it involves 3D printers or the Raspberry Pi, makers care about cool hardware, especially if it involves electronics.

But makers get nervous when it comes to motors. Pre-built quadcopters are growing in popularity, but I don't see many makers designing their own electronic speed controls (ESCs) or programming their own robotic arms. This is perfectly understandable. Motors are more complicated than other circuit elements. With motors, you don't just have to be concerned with electrical quantities such as voltage and current; you have to think about mechanical quantities such as torque and angular speed.

The topic of electric motors isn't easy, but the goal of this book is to make the concepts approachable to non-engineers. I assume a minimal background in mathematics and physics, and throughout the book, the emphasis is always on *making*. Instead of discussing the Lorentz force and electromagnetic flux, this book focuses on practical knowledge. Instead of bombarding you with equations, I'll show you the different types of motors available and the ways they can be controlled.

It takes time and patience to become comfortable with motors, but once you've ascended the learning curve, you'll be able to work on new and fascinating types of projects. Robots and remote-controlled vehicles will all fall within your grasp. The road is long, but I assure you that the destination is worth the journey.

Who This Book Is For

As the title should make clear, this is a book for makers. If you're looking for a textbook on phasor diagrams and Maxwell's equations, this isn't the book for you. If you're looking for practical information related to motor operation and control, you've come to the right place. If you want to know about the different types of motors and what they're good for, this is the book to have.

I've done my best to make motors comprehensible to non-engineers, but this book is not for beginners. In writing this book, I assume that you already know about volts, amps, and ohms. Further, I assume that you can look at a simple circuit diagram and get a sense for how the system works.

How This Book Is Organized

To present the topic of electric motors as clearly as possible, I've split the content into four parts:

- Part I, "Introduction," provides an overview of what motors are and how they work. Chapter 1, "Introduction to Electric Motors," introduces the history of electric motors and explains the two building blocks that make motor operation possible. Chapter 2, "Preliminary Concepts," expands on this, and explains how motors convert voltage and current into torque and angular speed.

- Part II, "Exploring Electric Motors," examines the many different types of motors available for makers. Specifically, the chapters in this part focus on DC motors, stepper motors, and servomotors. Later chapters investigate AC motors, linear motors, and gears. For each type of motor, the chapter explains how it operates and how it can be controlled.

- Part III, "Electrical Motors in Practice," presents three real-world applications of electric motors. Chapters 9 through 11 show how motors can be controlled with the popular circuit boards Arduino Mega, Raspberry Pi, and BeagleBone Black, respectively. Chapter 12, "Designing an Arduino-Based Electronic Speed Control (ESC)," explains how to build an electric speed control (ESC), and Chapter 13, "Designing a Quadcopter," explains how to build a quadcopter. The final chapter focuses on the important topic of electric vehicles.

- Part IV, "Appendixes," provides supplemental information that I hope will be helpful. Appendix A, "Electric Generators," discusses the topic of electric generators and the different types of machines that convert motion into electricity. Following that, the glossary in Appendix B provides definitions for many of the terms discussed throughout the book.

A handful of chapters present source code and circuit designs related to the content. These source code files and design files can be downloaded from http://motorsformakers.com.

Let Me Know What You Think

Feel free to email me at mattscar@gmail.com. I'm usually pretty good about responding promptly, though I won't promise a response to every concern.

INTRODUCTION TO ELECTRIC MOTORS

Of the many elements that can be placed in a circuit, none are as versatile or as exciting as the electric motor. Electric motors make it possible for robotic hands to grasp, electric cars to roll, and drones to fly. Quadcopters and 3D printers receive a great deal of attention, but to a system designer, they're just specialized motor control circuits.

In addition to being exciting, motors can also be hard to understand. When selecting a resistor, a designer only needs to be concerned with simple properties such as tolerance, temperature, and power rating. But when selecting a motor, there's a long list of questions that need to be addressed:

- Should the motor be direct current (DC) or alternating current (AC)?

- For a DC motor, should it be brushed or brushless?

- For a brushed DC motor, should it be a permanent magnet, series-wound, or shunt-wound motor?

- For a brushless DC motor, should it be an inrunner or an outrunner?

- Is the motor's Kv value sufficient for the system's speed and torque requirements?

- If the motor's torque is insufficient, what type of gears should be attached?

These aren't easy questions, and most books on electronics and robotics don't discuss them in depth. Instead, many books present specific circuits that require specific motors. They may mention why a particular motor is suitable for a task, but they don't provide enough information to enable you to select the right motor on your own.

This book takes a different approach. My goal is to present many different types of motors and show you how to select the right motor for your project. For each type of motor, I'll discuss its power requirements and the methods by which it can be controlled.

This book is aimed at makers, not scientists or engineers. As I discuss electric motors, I'll avoid the lengthy vector equations involving electric and magnetic fields. I studied this material in grad school, and I assure you that knowing the equations won't make your quadcopters faster or your remote-controlled cars more maneuverable.

1.1 Brief History

This book won't delve into the deep physics underlying motors, but there are two historical developments that every maker should know about. The first involves a moving needle in Denmark, and the second concerns a rotating wire in Hungary.

1.1.1 Oersted's Compass Needle

Hans Oersted was a Danish physicist who studied the relationship between electricity and magnetism. In 1820, he noticed something strange: Changing the current in a wire moved the needle of a nearby compass. Figure 1.1 shows what his experiment looked like.

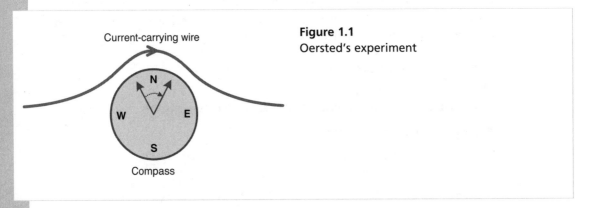

Figure 1.1
Oersted's experiment

Despite its simplicity, this experiment demonstrates the interaction between the two basic elements of an electric motor: changing current and a magnetic field. When these two components are in close proximity, the result is motion.

1.1.2 Jedlik's Self-Rotor

Oersted's experiment caused a flurry of activity in the scientist community. In France, Andre-Marie Ampere developed equations relating current in a wire to the magnetic field around the wire. In England, Michael Faraday devised a series of experiments that demonstrate how current-carrying wires move in the presence of a magnetic field.

But the credit for the first practical electric motor belongs to the Hungarian physicist Anyos Jedlik. Instead of placing the wire outside a compass, he wound it into coils and placed the coils inside a magnetic field. As current changes inside the coils, the coils rotate.

In 1827, Dr. Jedlik called his motor the electromagnetic self-rotor. Figure 1.2 shows what it looks like.

Figure 1.2
Jedlik's self-rotor

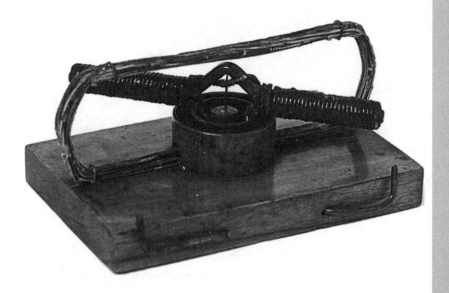

It has been nearly 200 years since Dr. Jedlik constructed his self-rotor, but today's rotary electric motors have essentially the same structure:

- The input electrical power is delivered through a current-carrying conductor.

- The current-carrying conductor is placed in the vicinity of a magnetic field.

Pretty simple, isn't it? The motors presented in this book have different shapes and configurations, but in nearly every case, the motion is produced by delivering current through a wire in the presence of a magnetic field

1.2 **Anatomy of a Motor**

Engineers like to be precise when describing their systems, and this is especially true for systems containing motors. The goal of this section is to introduce terminology for the different parts of a motor. These terms will be employed throughout this book.

Keep in mind that motors can be thought of as electrical elements or as mechanical elements. Therefore, the same part may have different names depending on whether the motor is considered electrically or mechanically.

1.2.1 External Structure

To describe the structure of an electrical motor, I'll start with the outside and proceed inward. Figure 1.3 presents a simple rotary motor.

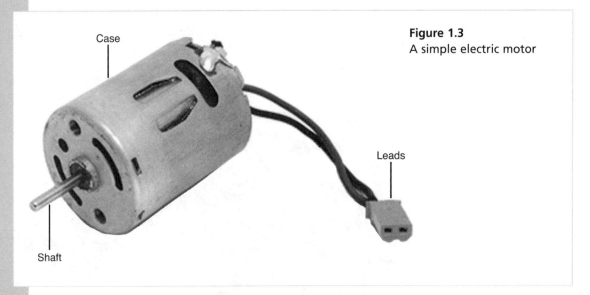

Case

Leads

Shaft

Figure 1.3
A simple electric motor

In describing the externals of an electric motor, three terms are commonly employed:

- **Case or shell**—The external housing surrounding the motor

- **Shaft**—The metal cylinder extending from the motor's center

- **Wires or leads**—The conductors carrying electricity to the motor

These terms should be straightforward to understand. The electrical input is delivered to the motor through its leads. As the motor operates, it rotates the shaft. This shaft is connected to a load such as the tire of an RC car.

1.2.2 Internal Structure

Figure 1.4 depicts a cross-section of a rotary motor. As current enters the motor, the central element rotates inside the case.

> **note**
> In certain types of motors, the shell rotates and the shaft remains fixed. One popular example is the outrunner brushless DC motor, which is discussed in Chapter 3, "DC Motors."

There are two ways to look at the motor's structure—mechanically and electrically. From a mechanical standpoint, the motor consists of two parts. The *rotor* is the part that moves, and the *stator* is the part that stays in place. The space separating the rotor and stator is called the *air gap*.

Viewed electrically, a motor's structure can be divided into another two parts. The *armature* is the part that receives current. In Figure 1.4, the motor's central element (the rotor) is the armature because it receives incoming current.

Figure 1.4
Internal structure of an electric motor

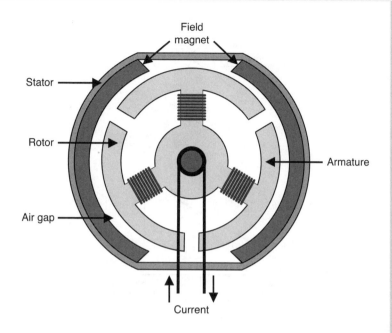

The second electrical part is responsible for generating the magnetic field. If the field is produced by permanent magnets, as it is in Figure 1.4, the second part is called the *field magnet*. If the magnetic field is produced by an electromagnet, the second part is called the *field winding*. Chapter 2, "Preliminary Concepts," explains how electromagnets work.

1.3 Overview of Electric Motors

When it comes to electric motors, there are a wide range of categories and subcategories to choose from. Chapters 3 through 8 discuss many of them in detail, and Figure 1.5 depicts a basic decision-making process that can be used to select which motor is suited for a task.

This diagram can help you make an initial assessment, but it's not a thorough breakdown of the many categories of electric motors. Later chapters will fill in the details. Also, some motors don't fit into this decision-making process. For example, universal motors (discussed in Chapter 6, "AC Motors") can operate on AC and DC power. Further, if any motor is connected to an encoder or position sensor, its angle can be measured and controlled.

One major decision that isn't mentioned in the flowchart involves the nature of the required motion. A motor that turns about an axis is a *rotary motor*. If it moves in a straight line, it's a *linear motor*. The vast majority of electrical motors are rotary, and Figure 1.5 applies to rotary motors only. But linear motors are important in many applications, especially robotics. These fascinating machines are discussed at length in Chapter 8, "Linear Motors."

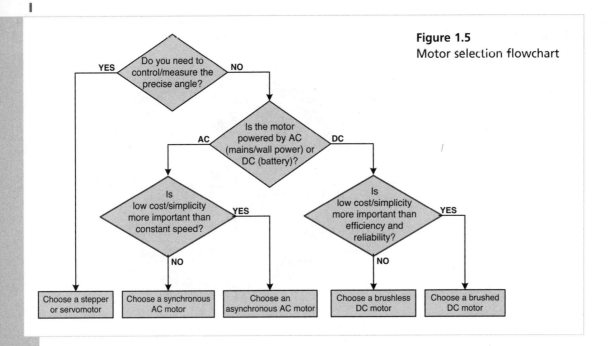

Figure 1.5
Motor selection flowchart

At a high level, electric motors can be categorized according to the nature of the input power. A DC motor receives DC (direct current) power, such as from a battery or power regulator. An AC motor receives AC (alternating current) power, such as from a wall socket.

1.3.1 DC Motors

DC motors accept DC electrical power, such as that provided by a battery. They're particularly common in maker projects. For example, every motor in a quadcopter or a remote-controlled car is a DC electric motor.

DC motors are divided into brushed and brushless motors. As will be explained in Chapter 3, the primary distinction between them involves the need for a commutator. Put simply, a commutator reverses voltage as the motor turns, thereby ensuring that the motor continues to turn. Motors with a mechanical commutator are called brushed motors or commutated motors. These motors are simple and inexpensive, but periodic maintenance is needed to keep them working properly.

Brushless DC motors, commonly called BLDCs, don't require maintenance as brushed motors do, but their structure is more complex. This means they cost more money and it takes significantly more effort to control them.

1.3.2 AC Motors

AC motors are common in industrial and household settings, and you'll find them in blenders, microwaves, and washing machines. AC motors come in two types: synchronous and asynchronous. The difference between them depends on how the motor's speed should be controlled. The speed of a synchronous motor is synchronized with the frequency of the incoming AC power.

But the majority of AC motors are asynchronous, which means their speed isn't synchronized with the frequency of the incoming power. These motors, frequently called induction motors, are popular, simple, and reliable. Chapter 6 discusses asynchronous and synchronous motors in detail.

1.4 Goals and Structure

This book has four goals:

- Present the different types of electric motors and their uses.

- Describe the circuits needed to control the different types of motors.

- Explain how to control motors with existing circuit boards.

- Show how to design real-world motor control circuits

Chapters 3 through 8 present many of the different types of electric motors. I'll introduce DC motors first, followed by stepper motors in Chapter 4, "Stepper Motors," and servomotors in Chapter 5, "Servomotors." Later chapters discuss AC motors (synchronous and asynchronous) and linear motors.

With regard to the second goal, the nature of a motor control circuit depends on the type of motor. That is, a circuit intended to control a stepper motor can't provide proper control of a servomotor, and you can't control a DC motor with an AC motor control circuit. Therefore, each chapter focusing on a motor's type also presents the basics of the motor's control. For example, the control circuitry needed for a stepper motor is discussed in Chapter 4, which presents stepper motors.

The third goal helps clarify theoretical issues related to motors and motor control. In the world of maker circuits, three of the most popular development boards are the Arduino Mega, the Raspberry Pi, and the BeagleBone Black. Chapters 9 through 11 take a close look at these boards and show how to program them to control different types of motors.

The last goal is particularly exciting. Chapters 12 through 14 get into the details of real-world motor control. Chapter 12, "Designing an Arduino-Based Electronic Speed Control (ESC)," explains how to design a fully functional ESC circuit, and Chapter 13, "Designing a Quadcopter," illustrates the full process of designing a quadcopter. Chapter 14, "Electric Vehicles," explains how electric motors are used in modern electric vehicles.

1.5 Summary

When it comes to electric motors, there are four levels of understanding:

- **Hobbyist**—"When I apply voltage and current, the motor's shaft turns."

- **Maker**—"The motor's shaft turns because the electromagnets in the stator are energized in sequence. The rotor's speed is proportional to voltage and the torque is proportional to current."

- **Engineer**—"The motor's impedance can be represented by the phasor $R_a + j\omega L_a$. If the input voltage is $V_m \sin(\omega t + 90°)$, the torque and speed can be computed as...."

- **Scientist**—"The electromagnetic tensor traveling through the conductor aligns the domains in the ferromagnetic material. This produces a magnetic vector field proportional to the material's permeability."

Engineers analyze motors mathematically and scientists are concerned with electromagnetic phenomena, but makers don't need the heavy math and physics. All a maker needs is an intuitive understanding of how motors work and how they can be controlled. This won't be sufficient for designing new motors, but when it comes to building systems with existing motors, this level of understanding is all you need.

With regard to these four levels, this book's goal is to help you advance from Level 1 (hobbyist) to Level 2 (maker). Many of the chapters focus on the different types of electric motors and the variety of methods available for controlling them. After reading this book, you should have a solid grasp of which motors are suitable for a given task. You should also be able to design control circuits without needing an existing recipe to fall back on.

This chapter began with a discussion of two scientific developments that led to the creation of modern electric motors. Hans Oersted discovered that a changing current deflects a compass needle. Anyos Jedlik went further and showed that coils of current-carrying wire in a magnetic field can serve as a practical motor. Many aspects of electrical motors have changed since Jedlik's time, but these motors still boil down to two parts: one component carries current (the armature) and the second produces the magnetic field (the field windings or field magnet).

After discussing the history of motors, this chapter explained aspects of their structure. On the outside, a motor has leads that deliver current and voltage to the motor and a shaft that delivers torque and speed to the load. If a motor is thought of as a mechanical element, it can be divided into the part that moves (rotor) and the part that doesn't (stator). As an electrical element, a motor can be split into the part that receives changing current (armature) and the part that produces the magnetic field (field magnet or field windings).

One reason motors make people nervous is that there are so many different types of them. Some require AC power, some require DC power, and universal motors can run on both kinds of power. Some motors are designed for motion control, whereas others specialize in torque and speed. This chapter has presented a brief overview of the different types, and later chapters will provide greater detail. The next chapter focuses on some basic properties that all motors share.

PRELIMINARY CONCEPTS

This book presents many types of rotary electric motors, each with a different purpose and structure. But viewed from a high level, these motors all perform the same operation: They convert voltage and current into torque and angular speed. In case you're not familiar with these quantities, this chapter begins by discussing torque and angular speed, and then explores the relationship between the two.

The rest of the chapter touches on three crucial subjects that relate to electric motors:

- Magnets

- Equivalent circuit elements for motors

- Power and efficiency

I'll keep the math to a minimum, but the terms and concepts involved can still be confusing. If you don't fully grasp the content of this chapter after the first reading, don't be alarmed. I recommend that you start by acquiring a basic familiarity with the material. Then, as the concepts become relevant in later chapters, you can return to this chapter to deepen your understanding.

2.1 Torque and Angular Speed

Force is one of the most fundamental quantities in physics and engineering, and most people have a basic understanding of it. The average person may not be comfortable with momentum or inertia, but everyone has an idea of what force is, especially the force of gravity.

Torque isn't as well understood. Some know that it's a kind of rotational force, yet torque and force have different units of measurement. Despite its relative unpopularity, torque is a quantity that every designer of motor systems needs to understand. The goal of this section is to make this clear and to show how a motor's torque relates to its angular speed.

2.1.1 Force

If an object undergoes any change in speed, whether it's acceleration (increase) or deceleration (decrease), the change is caused by a force. The amount of force can be obtained mathematically by multiplying the object's mass by its acceleration or deceleration.

For example, suppose I hold an object above the ground and let it fall. The object's change in motion is caused by the gravitational force acting on it. The amount of force equals its mass multiplied by the gravitational acceleration, 9.8 m/s². An object's gravitational force is referred to as its weight.

In America, weight is measured in pounds, and each pound consists of 16 ounces. Scientists and engineers measure weight and other forces in Newtons, abbreviated N. A force of 1 N is the force needed to accelerate a 1-kilogram mass by 1 meter per second per second (m/s²). One pound is approximately 4.45 Newtons, so my weight of 182 lbs. is close to 810 N.

Another fundamental aspect about force involves direction. A force's direction may change over time, but at any specific time, a force always acts in a straight line. When an object falls to the ground, it moves in the straight-line direction determined by the gravitational force.

2.1.2 Torque

Like force, torque is proportional to the object's mass and relates to an object's acceleration or deceleration. Unlike force, torque acts in a *circular arc*, not a straight line. When you turn a screw or twist the lid off a jar, you're exerting torque.

A good example of torque in action is an arm wrestling match. Figure 2.1 on the next page shows what this looks like.

In an arm wrestling match, each participant struggles to exert more torque than the other. If one succeeds, the difference in torque forces the opponent's arm to the table. The greater the difference in torque, the faster the opponent's arm will reach the table.

The situation is similar for rotary motors. A motor exerts torque through a shaft that connects to a load. If the motor can exert sufficient torque, the shaft will turn the load. If it can't exert enough torque, the shaft won't turn.

Earlier, I said that torque acts in a circular arc. An important difference between torque and force is that the torque depends on the arc's radius. More precisely, if the force is perpendicular to the arc's radius, the torque equals the force multiplied by the radius. Denoting torque as τ (tau), force as F, and the radius as r, the equation for torque is given as follows:

$$\tau = rF$$

This can be confusing, so let's take another look at the arm wrestling match. Figure 2.2 depicts the match with labels for force and radius.

Figure 2.1
A contest of torque

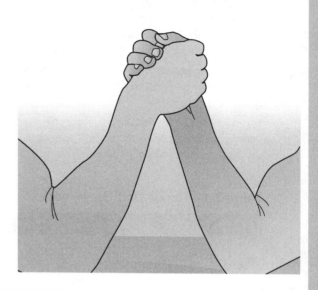

Figure 2.2
Force, radius, and torque

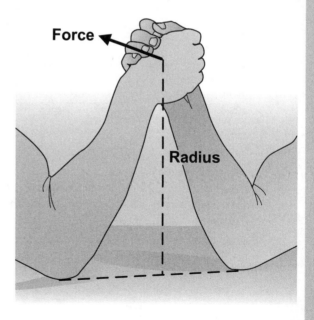

Consider the wrestler on the left. He's exerting force on his opponent's hand in order to turn it in a circular arc. In this case, the arc's radius is the line from the clasped hands to the table. The direction of the force changes over time, but it's always perpendicular to the radius. Therefore, the torque produced by the arm wrestler equals the force multiplied by the radius.

Scientists and engineers measure torque in Newton-meters, abbreviated N-m, or milliNewton-meters, abbreviated mN-m. Occasionally, torque will be given in Newton-centimeters, abbreviated N-cm.

For American motors, data sheets commonly use Imperial units, such as pound-force-feet (lb-ft), pound-force-inches (lb-in), or ounce-force-inches (oz-in). Table 2.1 provides conversion values between common units of torque.

Table 2.1   Torque Unit Conversions

From	To	Multiply By
oz-in	N-m	0.007062
oz-in	lb-ft	0.005208
oz-in	lb-in	0.0625
lb-ft	oz-in	192.0
lb-ft	N-m	1.356
lb-ft	lb-in	12.00
N-m	oz-in	141.6
N-m	lb-ft	0.7376
N-m	lb-in	8.851
lb-in	oz-in	16
lb-in	N-m	0.1130
lb-in	lb-ft	0.08333

The torque exerted by a motor depends in large part on the nature of its load. To see what I mean, imagine that you're taking part in an arm wrestling match. If your opponent is much weaker than you are, you can win the match quickly without exerting yourself. This is the *no-load condition*. When a motor is in the no-load condition, there's no load—the shaft rotates quickly and exerts minimal torque.

Now suppose your arm wrestling opponent is stronger than you are. No matter how much you exert yourself, your opponent's hand won't budge. This is the *stall condition*. When a motor is in the stall condition, it exerts a great deal of torque, but the rotational speed is zero because the load is too great.

2.1.3 Angular Speed

Speed measures how quickly an object moves. In other words, speed tells you how far an object travels in a given amount of time. Angular speed is similar, but identifies the angle an object rotates through in a given amount of time.

When I took engineering courses in college, angles were always measured in radians and angular speed was measured in radians per second. But for electric motors, angle is measured in degrees (°) and angular speed is measured in revolutions per minute, or RPM. A motor's rotational speed is denoted as ω (omega). If an object rotates with a speed of 12°/sec, the speed in RPM can be computed as follows:

$$\omega = \left(\frac{12\,\text{deg}}{\text{sec}}\right) \cdot \left(\frac{60\,\text{sec}}{1\,\text{min}}\right) \cdot \left(\frac{1\,\text{rev}}{360\,\text{deg}}\right) = \frac{2\,\text{revs}}{\text{min}} = 2\ \text{RPM}$$

It's important to clearly understand the difference between torque and angular speed. Angular speed tells you how many revolutions an object completes per minute. Torque tells you how much force the object can exert as it rotates. An object can rotate at high speed with low torque, or with high torque at low speed.

2.1.4 The Torque-Speed Curve

If a motor's shaft isn't connected to a load, the motor's speed is referred to as its *no-load speed*, denoted ω_n. For most motors, this is the maximum speed.

If the motor's load is so large that its shaft can't turn, the motor's torque is called its *stall torque*, denoted τ_s. This is the maximum amount of torque that the motor is capable of exerting.

When selecting a motor for a project, it's a good idea to know what these values are. For example, one brushed DC motor sold on pololu.com has the following characteristics:

- ω_n equals 17,000 RPM for an input voltage of 3 V.

- τ_s equals 0.75 oz-in for a stall current of 3.85 A.

In an ideal world, in addition to providing ω_n and τ_s, manufacturers would tell us how the motor behaves between those extreme states. That is, they'd tell us the motor's torque and speed when the load is greater than zero but less than the maximum supportable limit.

The relationship between a motor's torque and speed can be illustrated graphically with a line called the *torque-speed curve*. Most manufacturers don't provide torque-speed curves for their motors, but they can be obtained experimentally with the right measuring equipment, such as a *dynamometer*. Figure 2.3 depicts a simplistic example of a torque-speed curve.

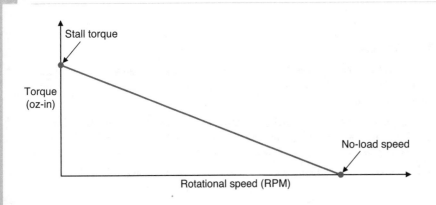

Figure 2.3
An example of torque-speed curve

The shape of the torque-speed curve depends in large part on the motor's type. Also, a curve is only valid for a given voltage and current. That is, ω_n and τ_s change as the input electrical power is increased or decreased.

2.2 Magnets

Whether you're selecting a motor for a new project or designing a control circuit, it helps to know how the motor works. I'm not talking about the complex physical laws, but the basic manner in which it converts electricity into rotary motion. This conversion depends on the interaction of magnets, so let's review three basic facts:

- Every magnet has two poles: north (N) and south (S).

- Opposite poles attract and like poles repel. In other words, when magnets come close to one another, they orient themselves to bring opposite poles closer and similar poles further apart.

- Different magnets have different strengths. The stronger the magnet, the more its poles attract/repel the poles of other magnets.

If you're not already familiar with these points, you can verify them easily by experimenting with bar magnets.

There are two types of magnets, and both types are common in the world of electric motors:

- **Permanent magnets**—When certain materials are brought near high electric current, such as lightning, they acquire permanent magnetic behavior.

- **Electromagnets**—When a current-carrying wire is wrapped into a coil, it behaves like a magnet. This behavior is temporary, and stops when current is removed. If the coil is wrapped around an iron core, the magnetic behavior grows stronger.

Permanent magnets don't need power to operate and they're generally strong. Unfortunately, their poles are always fixed in the same positions and they tend to be expensive.

The main advantage of electromagnets is that their magnetic behavior can be controlled by changing the current. That is, altering the power can alter an electromagnet's strength and the location of its poles. The main disadvantage is power consumption—it takes a lot of current to make an electromagnet as strong as a permanent magnet.

The ability to control the magnetic behavior of electromagnets is central to the operation of electric motors. Therefore, it's helpful (but not necessary) to know how an electromagnet's current determines the locations of its north and south poles.

Let's refer to the side where the current enters as the *top* of the electromagnet and the side where the current exits as the *bottom*. If you look downward at the top of an electromagnet, you'll see that the current flows in a clockwise or counterclockwise orientation. This orientation determines the electromagnet's pole locations, or its *polarity*. This is set by two rules:

- If current flows in a counterclockwise direction, the top is the north pole and the bottom is the south pole.

- If current flows in a clockwise direction, the bottom is the north pole and the top is the south pole.

Figure 2.4 presents two electromagnets and their north and south poles. Note that, if the direction of current is reversed, the north and south poles switch positions. Put another way, reversing the current reverses the electromagnet's polarity.

Figure 2.4
Electromagnets and their poles

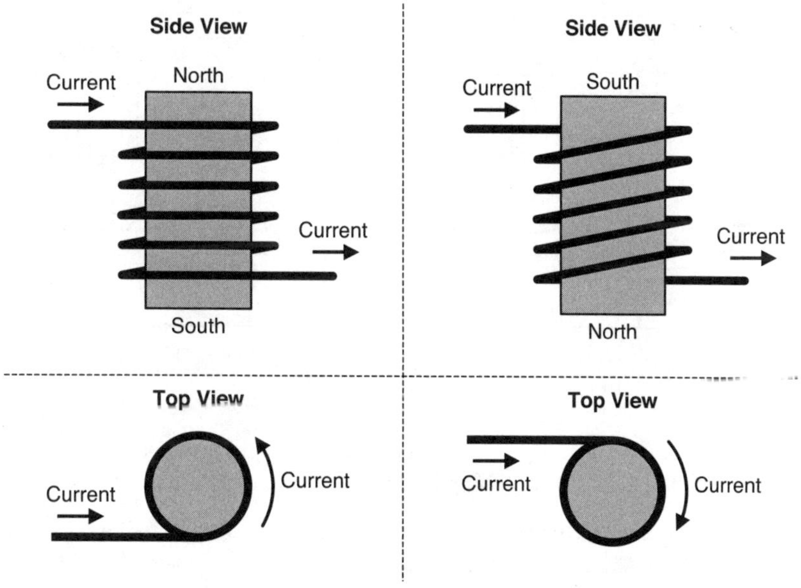

Now let's consider an experiment involving four electromagnets, named A, A', B, and B'. The A and A' electromagnets are connected to the same wire, and so are B and B'. In the center, a permanent magnet, called a *bar magnet*, is free to rotate. Figure 2.5 illustrates the positions of the electromagnets and bar magnet.

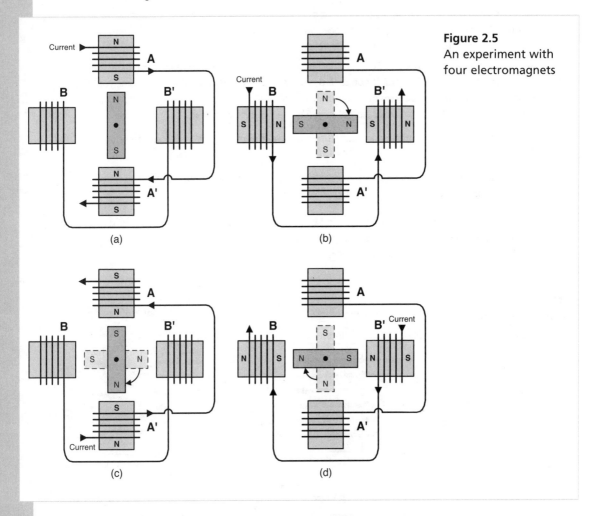

Figure 2.5
An experiment with four electromagnets

When current is applied to one pair of electromagnets, the bar magnet rotates to align itself with the electromagnets' poles. As depicted in the figure, there are four possibilities:

- In Figure 2.5a, current flows from A to A'. The south pole of A faces the magnet and the north pole of A' faces the magnet. The bar magnet aligns itself so that its north pole faces A and its south pole faces A'.

- In Figure 2.5b, current flows from B to B'. The north pole of B faces the magnet and the south pole of B' faces the magnet. The bar magnet rotates so that its north pole faces B' and its south pole faces B.

- In Figure 2.5c, current flows from A' to A. The north pole of A faces the magnet and the south pole of A' faces the magnet. The bar magnet rotates so that its north pole faces A' and its south pole faces A.

- In Figure 2.5d, current flows from B' to B. The south pole of B faces the magnet and the north pole of B' faces the magnet. The bar magnet rotates so that its north pole faces B and its south pole faces B'.

If current is delivered to the electromagnets in this sequence, the bar will rotate 360° in the counterclockwise direction. If current is delivered in the reverse sequence, the bar will rotate 360° in the clockwise direction.

I recommend that you study Figure 2.5 until you're comfortable with it. This process of converting electrical power (current through the electromagnets) into mechanical power (rotation of the bar magnet) is essentially the same process used by most of the motors in this book.

2.3 Equivalent Circuit Element

When you design circuits with motors, it's important to know how the motor behaves electrically. Does it act like a resistor, a capacitor, an inductor, or a diode? Or does it resemble a combination of these?

Given the diversity of electric motors, it's impossible to construct a single model that fits all types. But it is possible to account for electrical characteristics that apply to a wide range of motors. In particular, this section looks at electrical losses that are common to motors and then discusses the phenomenon of back-EMF.

2.3.1 Electrical Losses

When a motor rotates with no load, it still draws current. This is called the motor's *no-load current*, denoted I_o. This is needed to magnetize the iron cores of the electromagnets, so the electrical loss is called *iron loss*. When you're analyzing a circuit, this loss is accounted for by subtracting I_o from the current entering the motor.

Another source of electrical loss in a motor is called *copper loss*. This relates to the resistance of the armature, which is the portion of the motor that receives electrical power. An armature's resistance is denoted R_a, and if I is the current entering the motor, Ohm's law tells us that the voltage loss across the armature is $(I - I_o)R_a$. Therefore, if V is the motor's input voltage, the voltage that contributes to the motor's operation equals $V - (I - I_o)R_a$.

In addition to resistance, a motor's armature adds inductance. This inductance, denoted L_a, is usually so small that it doesn't need to be considered. But the armature's inductance is proportional to the frequency of the current flowing through it. Therefore, if the motor is part of a high-frequency circuit, L_a can have a significant effect.

Figure 2.6 presents an equivalent circuit for a motor that takes these losses into account.

Most electromagnets have iron cores, but not all do. Motors without iron cores are called *coreless* or *air-core* motors. Coreless motors are weaker than motors with iron cores, but they're lighter and have lower electrical losses.

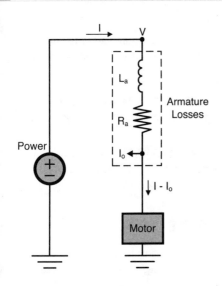

Figure 2.6
An equivalent circuit for armature losses

2.3.2 Back-EMF

Electric motors have a fascinating property that every maker should be aware of. As a motor rotates, the interaction of its conductors and magnets generates a voltage proportional to the speed of rotation. If this rotation is caused by an external force, such as the flowing of a waterfall, this voltage can be used to provide power. In this case, the motor behaves as a generator, which is the topic of Appendix A, "Electric Generators."

But if the motor rotates in response to electrical power, the generated voltage opposes the incoming current. In this case, the voltage is referred to as *back-EMF*, where EMF stands for electromotive force, which is a synonym for voltage. This phenomenon may also be referred to as *counter-EMF*, or *cemf*. Back-EMF increases as the voltage across the motor increases, but it's always less than the motor's input voltage.

The shape of the back-EMF depends on the type of the motor and the power delivered to it. But in all cases, the back-EMF can be modeled as a voltage source that opposes the direction of the incoming current. This is shown in Figure 2.7, which presents a more complete circuit model for an electric motor.

Back-EMF is particularly important to understand when you're designing circuits for brushless DC motors. To control these motors, the controller needs to know the motor's shaft angle, and it figures out this angle by determining the motor's back-EMF. Chapter 3, "DC Motors," discusses brushless DC motors and describes different methods of using back-EMF.

Figure 2.7
An equivalent circuit for an electric motor

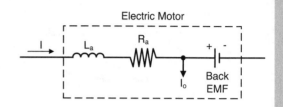

2.4 Power and Efficiency

An electric motor may be able to operate over a range of speed and torque, but every motor has an operating point at which it's particularly productive. At this *peak efficiency point*, the motor converts electrical power to rotational power with an absolute minimum of wasted power.

When you're selecting a motor for an application, efficiency is a major concern. But before I can explain how efficiency is computed, I need to introduce the concept of power the way scientists and engineers understand it. Before I can introduce the concept of power, I need to discuss the concept of work.

2.4.1 Work

Work is performed whenever a force acts on an object and moves it over a distance. If the force and distance are parallel, the work is proportional to both the force and the distance. That is, if a force, F, moves an object by a distance of d in the force's direction, the work exerted is Fd.

For example, if an object weighing 20 N falls a distance of 2 meters, the work performed by the gravitational force is 40 N-m. If the object doesn't move or moves in a direction perpendicular to the force, the work performed by the force is zero.

Rotational work is similar, but involves torque and angle. If a torque, τ, turns an object through an angle, θ, the work performed equals $\tau\theta$. As an example, if you exert a constant torque of τ as you twist the lid off a jar, and the lid turns an angle of θ, the amount of work you've performed is $\tau\theta$.

2.4.2 Rotational Power

Power is the rate at which work is performed. If work is constant over an interval of time, the power is computed by dividing work by the time interval. Denoting power as P, work as W, and the time as t, this relationship can be expressed mathematically:

$$P = \frac{W}{t}$$

For example, if two electric screwdrivers perform equal amounts of work, but the first finishes in half the time, we say that on average, the first screwdriver used twice as much power as the second. Similarly, if one car engine can accelerate from low speed to high speed in less time than another engine, we say that it's more powerful.

Power is measured in watts, abbreviated W, and 1 W equals 1 N-m/s. Another common unit is *horsepower*, or hp; 1 hp equals 745.7 W, and 1 W equals 0.00134 hp. For DC motors, power is commonly expressed in watts. For AC motors, power is generally expressed in horsepower.

As mentioned earlier, rotational work equals torque multiplied by the angle through which the torque is exerted. Rotational power equals torque times the angle divided by time. This is the same thing as torque times angular speed. Therefore, if torque is given in N-m and angular speed is given in RPM, the mechanical power in watts can be computed with the following equation:

$$P_{mechanical} = 0.1047\tau\omega$$

If the angular speed is given in RPM and torque is given in oz-in, the formula for rotational power is given as follows:

$$P_{mechanical} = 0.0007396\tau\omega$$

2.4.3 Electrical Power

Electrical power is even easier to understand than rotational power: Power equals voltage times current. Denoting voltage as V, current as I, and the power as P, the equation is given as follows:

$$P_{electrical} = VI$$

This assumes that voltage is measured in volts, current is measured in amperes, and power is measured in watts. If 2 A of current flows at 5 V, the electrical power equals 10 W.

As discussed earlier, the voltage loss due to the armature's resistance is IR_m and the current loss due to iron loss is I_o. Taking these losses into account, the power reaching the motor can be given as follows:

$$P_{motor} = (V - IR_m)(I - I_o)$$

This expression identifies the electrical power that the motor successfully converts to mechanical power. The mechanical power equals torque, τ, multiplied by speed, ω. Equating electrical and mechanical power, we obtain an important equation:

$$(V - IR_m)(I - I_o) = \tau\omega = P_{mechanical}$$

2.4.4 Efficiency

No electric motor converts all of its incoming electrical power to mechanical power. The ratio between the motor's input power and output power is called *efficiency*, denoted as η. This value can be obtained with the following equation:

$$\eta = \frac{P_{output}}{P_{input}}$$

An ideal electric motor has an efficiency of 1, but real-world motors generally have efficiencies between 0.5 and 0.9. For example, if a motor produces 1.5 watts of mechanical power for every 2.0 watts of input electric power, its efficiency is 1.5/2.0 = 0.75.

For an electric motor, efficiency can be computed in the following way:

$$\eta = \frac{P_{output}}{P_{input}} = \frac{P_{mechanical}}{P_{electrical}} = \frac{\tau\omega}{VI}$$

Using a relationship derived earlier, the numerator can be expressed in terms of voltage, current, and losses:

$$\eta = \frac{P_{mechanical}}{P_{electrical}} = \frac{(V - IR_m)(I - I_o)}{VI} = \left(\frac{V - IR_m}{V}\right)\left(\frac{I - I_o}{I}\right)$$

Efficiency increases as V increases and as R_m decreases. Therefore, so long as the voltage is kept within the motor's limits, efficiency can be increased by increasing voltage. This explains why high-power motors running at their rated speed are more efficient than low-power motors running at their rated speed.

The relationship between efficiency and current is more complex. On one hand, an increase in current increases the copper loss (IR_m), so this reduces efficiency. On the other hand, an increase in current decreases the effect of the iron loss (I_o), so this increases efficiency.

Many vendors specify a motor's peak efficiency parameters, and it's a good idea to keep them in mind. You should always choose a motor whose peak efficiency characteristics closely match those of the desired workload. That is, the motor's torque and speed at peak efficiency should be close to the torque and speed that it's intended to provide.

The penalty for motor inefficiency is more than just wasted power. The copper loss in the armature produces heat equal to I^2R_m. If this heat grows too large, it can damage the motor and the surrounding circuitry.

2.5 Summary

When you're selecting a motor, it's important to clearly understand the specifications provided by the manufacturer. These specifications center on four physical quantities: current, voltage, speed, and torque. Most of these quantities should be clear, but the first part of this chapter discusses the nature of torque and rotational speed.

When analyzing a motor's behavior, we concern ourselves with two extreme conditions: the no-load condition and the stall condition. The no-load condition arises when the motor rotates quickly without a load. The stall condition occurs when the motor fully exerts itself but fails to move the load. These conditions play a central role in determining a motor's torque-speed curve.

To see how electric motors convert electric power to mechanical power, it's crucial to understand magnets and electromagnets. An electromagnet is a coil of wire that usually surrounds an iron core. Unlike permanent magnets, whose magnetic properties are fixed, the behavior of an electromagnet can be altered by changing the current flowing through the wire. The drawback to using electromagnets is that they require a great deal of current.

When you connect a motor to an electrical circuit, it helps to understand the motor's electrical characteristics. Two important characteristics are iron loss and copper loss. Iron loss is the current needed to energize the motor's electromagnets. Copper loss is the resistive heating (I^2R) within the armature.

Once you have a general idea of the type of motor you're interested in, the next step is to select the motor whose characteristics are best suited to the task. To keep wasted power to a minimum, the motor's peak efficiency should be reached during normal operation. Efficiency is the ratio of mechanical power to electrical power, and the higher the efficiency, the more input power is used productively.

DC MOTORS

If your gadget is intended to be powered by batteries, solar power, a USB interface, or any other source of direct current (DC), the only motors you should consider are DC motors. These are straightforward to use and understand, and if you grasped the content of Chapter 2, "Preliminary Concepts," you should have no trouble working with them.

When you're selecting the right DC motor for your application, the first question to ask is whether it should be brushed or brushless. This is the primary distinction among DC motors, and a large part of this chapter discusses motors of both types. Due to their greater complexity, brushless motors will be examined in greater length.

The first part of this chapter discusses topics that apply to all DC motors. This section focuses on the relationship between a DC motor's torque and current and the relationship between its speed and voltage. This discussion also presents high-level principles behind DC motor control.

The chapter ends with a brief overview of batteries. When it comes to voltage and current, motor control circuits have specific needs that require special types of batteries. Motor circuits rely primarily on rechargeable batteries, and the last part of the chapter explains the pros and cons of different battery types.

3.1 DC Motor Fundamentals

Brushed and brushless DC motors have different internal structures and different methods of control, but they have four characteristics in common:

- Torque is approximately proportional to current.

- Speed is approximately proportional to voltage.

- Control circuitry employs electrical switches to deliver power to the motor.

- A controller can govern the motor's operation using PWM (pulse width modulation) signals.

This section discusses each of these topics. Later sections discuss the two types of motors separately.

3.1.1 Torque, Current, and K_T

Ampere's Force Law tells us how much force a magnetic field exerts on a current-carrying wire. This is a complex equation that involves vectors and calculus, so I'll summarize: As the current entering the armature of a motor increases, the motor's torque increases.

Using a digital torque meter, I measured how much current a brushed DC motor draws as its load increases. Figure 3.1 illustrates the relationship between a motor's load and the incoming current.

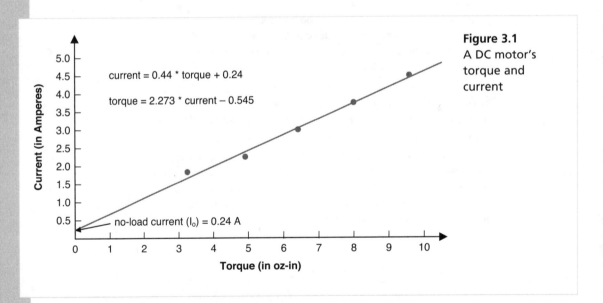

Figure 3.1
A DC motor's torque and current

For DC motors, the relationship between torque and current can be approximated with a straight line. This means that the ratio between torque and current is generally constant. This constant is referred to as K_T, and many datasheets provide this value in ounce-inches/ampere (oz-in/amp), pound-inches/ampere (lb-in/amp), Newton-meters/ampere (N-m/A), or Newton-centimeters/ampere (N-cm/A). In Figure 3.1, K_T equals 2.273 oz-in/amp.

As discussed in Chapter 2, when a motor's shaft rotates but exerts no torque, it's in the no-load condition. The current drawn by a motor in its no-load condition is called the *no-load current*, denoted as I_o. In Figure 3.1, I_o equals 0.24 A.

I_o is the minimum amount of current needed to put the motor in motion. Therefore, if the motor's armature receives a current of I, the torque produced by the motor equals $K_T(I - I_o)$. If the torque is given in N-m and K_T is given in oz-in, the relationship is given as follows:

$$\tau = K_T\left(I - I_o\right) \cdot \left(\frac{0.278\ \text{N}}{1\ \text{oz}}\right) \cdot \left(\frac{0.0254\ \text{m}}{1\ \text{in}}\right)$$
$$= 0.0070612\ K_T\left(I - I_o\right)$$

3.1.2 Rotational Speed, Voltage, and K_V

Just as a motor's torque increases with current, its rotational speed increases with voltage. This angular speed, denoted as ω, is given in rotations per minute, or RPM. Using a tachometer, I measured the speed of a brushed DC motor at different levels of voltage. Figure 3.2 presents the results.

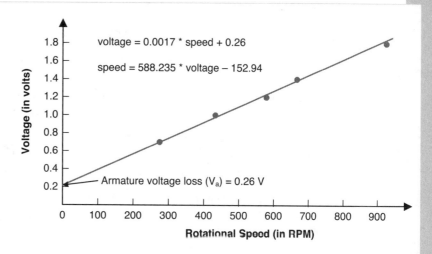

Figure 3.2
Speed versus voltage

voltage = 0.0017 * speed + 0.26

speed = 588.235 * voltage − 152.94

Voltage (in volts)

Armature voltage loss (V_a) = 0.26 V

Rotational Speed (in RPM)

As with the torque-current graph, the voltage-speed graph closely resembles a straight line. The constant that specifies a motor's speed/voltage ratio is called K_V, and this is usually given in units of RPM/V (revolutions per minute per volt). In Figure 3.2, K_V equals 588.235 RPM/V.

As discussed in Chapter 2, every motor's armature has resistance, R_a. Some of the voltage entering the motor will drop across R_a, and we refer to this voltage loss as V_a. In Figure 3.2, V_a equals 0.26 V.

If V is the total voltage applied, the motor's speed equals $K_V(V - V_a)$. Denoting the motor's current as $I - I_o$, it's clear that $V_a = (I - I_o)R_a$.

3.1.3 The K$_T$-K$_V$ Tradeoff

The last part of Chapter 2 explained that motors convert input voltage (V − V$_a$) and input current (I − I$_o$) to torque (τ) and speed (ω). This relationship can be expressed by equating the input electrical power to the output mechanical power, which leads to the following equation:

$$(V - V_a)(I - I_o) = \tau\omega$$

You've seen how K$_T$ relates to a motor's torque and how K$_V$ relates to a motor's speed. If K$_T$ is given in oz-in/A and K$_V$ is given in RPM, the resulting equation is given as follows:

$$(V - V_a)(I - I_o) = \left\{0.0070612\, K_T\left(I - I_o\right)\right\} \cdot$$
$$\left\{K_V\left(V - V_a\right) \cdot \left(\frac{1\,\text{min}}{60\,\text{sec}}\right) \cdot \left(\frac{2\pi}{\text{revolution}}\right)\right\}$$

Combining the constants and dividing both sides by (V − V$_a$)(I − I$_o$) leads to the following result:

$$1 = 0.000739447\, K_T K_V$$

This can be simplified to produce the final equation:

$$K_T K_V = 1352.36$$

This result makes it possible to draw three conclusions about electric motors:

- If you know one constant, it's easy to compute the other. For this reason, most datasheets only provide K$_V$.

- No motor excels at converting current to torque (high K$_T$) *and* converting voltage to speed (high K$_V$). If one value is relatively high, the other must be relatively low.

- If a motor's purpose is to run quickly, select a motor with high K$_V$ and low K$_T$. If its purpose is to provide torque, select a motor with high K$_T$ and low K$_V$.

For most applications, the average motor provides too much speed and too little torque. That is, if you connect an average brushed motor to a quadcopter's propeller or a robotic arm, the motor's torque won't be sufficient to turn the shaft.

For this reason, many systems insert gears between the motor and the load. Gears make it possible to increase torque and reduce speed. This topic is discussed at length in Chapter 7, "Gears and Gearmotors."

 note

The torque/current and speed/voltage relationships are just as valid for AC motors as they are for DC motors. But DC motor speed is controlled primarily by increasing voltage, and AC motor speed is controlled primarily by increasing the frequency of the incoming power. This is why datasheets for DC motors specify K$_V$ and datasheets for AC motors don't.

3.1.4 Switching Circuitry

The preceding chapter explained that the main disadvantage of using electromagnets is power consumption. This is why even small motors may need tens of amps to function properly.

The circuitry that governs a motor's operation is called the *controller*. In modern systems, this is an integrated circuit. In most maker-focused devices, this is a microcontroller or a low-power processor. These devices run on milliamps, so they can't directly provide a motor with the power it needs. For this reason, motor circuits need electrical switches.

Electrical Switches

In an electric circuit, a mechanical switch creates a conductive path when a button is pressed. An electric switch works in essentially the same way. When the controller applies a small voltage to one terminal, the switch creates a conductive path that allows current to flow through the other terminals.

The left side of Figure 3.3 presents the standard two-terminal symbol for a mechanical switch. The right side presents a three-terminal symbol I created to represent an ideal electrical switch. When the input voltage is greater than 0 V, the switch closes and creates a conductive path between the other two terminals. If the input voltage is zero or less, the switch is open and no current flows between the other two terminals.

Figure 3.3
A mechanical switch and an electrical switch

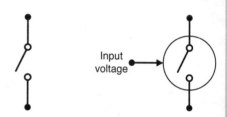

An example will show how this electrical switch can be used. Suppose that a 3.3 V microcontroller is employed to control a brushed DC motor. This chip can't deliver power directly to the motor, but if power is connected through an electric switch, the controller can govern the motor's current by turning the switch on and off. Figure 3.4 shows what this looks like

When $V_{CONTROLLER}$ is zero, the switch is open. No current flows, so the motor doesn't turn. But when $V_{CONTROLLER}$ is greater than zero, the switch closes and delivers current from V_{POWER} through the motor, causing it to rotate.

Transistors as Electrical Switches

Ideal electrical switches don't exist in reality, but we can approximate such a switch with a device called a *transistor*. To be specific, most modern motor circuits rely on metal-oxide-semiconductor field-effect transistors (MOSFETs) or insulated-gate bipolar transistors (IGBTs) to serve as switches.

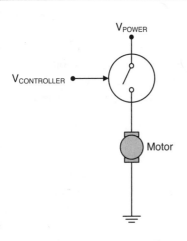

Figure 3.4
Controlling a motor with an electric switch

Don't be intimidated by the confusing names. Both types of devices serve the same purpose as the electrical switch in Figure 3.4. Both have three terminals—one that receives voltage from the controller and two that pass current when the controller's voltage is high. Figure 3.5 shows what their circuit symbols look like.

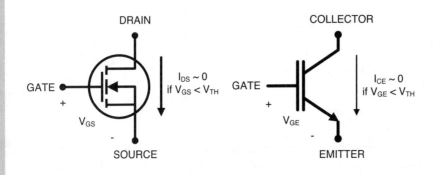

Figure 3.5
Real-world electric switches: a MOSFET and an IGBT

As shown in the figure, the input terminal of a MOSFET or IGBT is referred to as the gate. For a MOSFET, the terminals that deliver current are the drain and source. For an IGBT, the current-carrying terminals are the collector and emitter.

Let's review the properties of an ideal electric switch. When the input voltage is less than or equal to zero, the resistance between the other two terminals is infinite. When the input voltage is greater than zero, the resistance between the two terminals is zero.

MOSFETs and IGBTs come close to this ideal, but there are three non-ideal aspects of their behavior to be aware of:

- To close the switch, the voltage between the gate and lower terminal must be greater than a threshold voltage, denoted as V_{TH}. For a MOSFET, V_{TH} is typically between 0.5 and 1 V. For an IGBT, V_{TH} is commonly between 3 and 8 V.

- When the gate voltage is less than V_{TH}, the current allowed to pass through the other two terminals is so small as to be considered practically zero.

- When the gate voltage is greater than or equal to V_{TH}, the resistance between the other terminals is low, but it's not zero. For a MOSFET, the resistance between the drain and source, $R_{DS(on)}$, is as low as 0.03 Ω. For an IGBT, the voltage-current relationship isn't linear, but the voltage drop between the two terminals is less than that between the terminals of a comparable MOSFET.

MOSFETs and IGBTs have similar purposes, terminals, and operating characteristics, but MOSFETs can switch current on/off more quickly and are generally less expensive. In contrast, IGBTs can switch greater amounts of current, and the voltage drop from collector to emitter is less than the drain-source voltage drop of a similarly capable MOSFET.

As a rule of thumb, MOSFETs are better suited for circuits with small- to medium-sized motors. IGBTs are better suited for circuits with large motors. When working with either type of transistor, be sure to read the datasheet to ensure that its characteristics are suitable for the circuit.

Maker circuits tend to be focused on small- to medium-sized motors. Therefore, the rest of this book will rely exclusively on MOSFETs to serve as electric switches.

3.1.5 Pulse Width Modulation (PWM)

With an electrical switch, the controller can turn a motor's current fully on or fully off. But what if you want the motor to rotate at 75% of its full speed? What if you want the motor's speed to ramp up gradually? Increasing the controller's voltage won't help—once the gate voltage exceeds the transistor's threshold voltage, increasing the gate voltage further won't substantially increase the current.

Instead, controllers govern the motor's behavior by delivering pulses that open and close the switch for precise amounts of time. This pulse delivery is referred to as pulse width modulation, or PWM.

The concept underlying PWM is simple. The controller delivers a series of pulses to the gate of the switch. These pulses open and close the switch and the switch delivers pulses of current to the motor. The controller generates pulses at equal intervals, so the wider the pulse, the more current reaches the motor and the faster it runs.

Figure 3.6 shows what a PWM pulse train looks like. In this case, the controller delivers four pulses to the switch's gate.

As labeled in the figure, T is the time between the rising edges of two adjacent pulses. This interval is referred to as the *frame* or the *period*. The controller sets T by configuring the PWM frequency, which equals 1/T.

t is the length of time that the controller's signal is high (greater than the threshold voltage). This time is referred to as the *pulse width*. The *duty cycle* is the ratio of the pulse width to the frame, or t/T. In Figure 3.6, the duty cycle is 0.3.

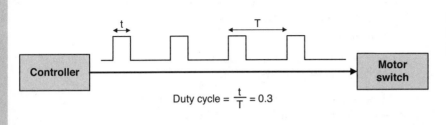

Figure 3.6
Pulse width modulation (PWM)

For example, if the controller sets the PWM frequency to 500 Hz, T = 1/500 seconds = 2 ms. If the duty cycle is 0.4, each pulse occupies 40% of the frame, which means t = (2 ms)(0.4) = 0.8 ms. If the duty cycle is 1.0, the switch remains fully closed and a maximum amount of power reaches the motor.

Choosing the right PWM frequency is critical. In making this decision, we have two important factors to consider:

- If the frequency is too low, the rise/fall of the power reaching the motor will cause it to rotate in a rough, jerky fashion.

- If the frequency is too high, the pulses will be too narrow to open and close the switches properly. In addition, the electromagnets will generate heat, decreasing the motor's efficiency.

There are no clear rules regarding PWM frequency, but many servomotor circuits directed at hobbyists expect a frequency of 50 Hz. This corresponds to a frame of 20 ms. The best place to look for information is the datasheet, but if the frequency isn't given, it's safe to assume that the PWM frequency is 50 Hz.

Frequencies between 30 Hz and 20 kHz produce noise within the human range of hearing. If this is a concern, you may want to set the PWM frequency higher than 20 kHz.

3.2 Brushed Motors

This section puts aside general motor theory and starts looking at actual motors. The best place to start is brushed motors, which are the simplest practical motors discussed in this book. Their internal structures are simple and they're easy to control. As an example, Figure 3.7 shows what a 12-volt brushed DC motor looks like.

Despite their simplicity, brushed DC motors have features that may make them unsuitable for certain applications. To understand why this is the case, you need to be familiar with the idea of mechanical commutation.

3.2.1 Mechanical Commutation

As discussed in Chapter 1, "Introduction to Electric Motors," every electric motor contains two parts: current in a conductor and a magnetic field. It's important to note that the current must change over time. If the current in the conductor is constant, the motor won't make a complete rotation.

Figure 3.7
A 12-volt brushed motor

This may seem odd because this chapter focuses on DC motors and DC implies constant current. To help make this clear, Figure 3.8 presents a loop of wire carrying current between two magnets. The wire's orientation changes in each case, but the current, I, is the same.

Figure 3.8
A rotating loop carrying constant current in a magnetic field

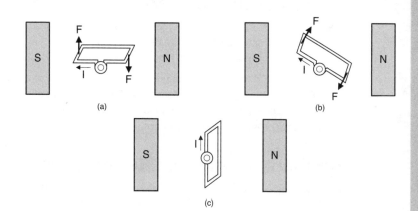

The arrows labeled F represent the forces acting on the wire. These forces depend on three things: the wire's orientation (horizontal, vertical, at an angle), the current's direction, and the direction of the magnetic field (from N to S). In Figure 3.8, the three situations are explained as follows:

- In Figure 3.8a, the wire is horizontal. The current enters its left side and leaves through its right. Because of the wire's position and the direction of the current, two forces act on it: one that pushes the wire's left side upward and one that pushes the right side downward. As a result, the wire rotates in a clockwise orientation.

- In Figure 3.8b, the wire is positioned at an angle. Again, two forces are produced, but now their directions are different. The upper-left part of the wire is pushed upward and to the right and the lower-right part is pushed downward and to the left. As a result of these forces, the wire continues rotating in a clockwise orientation.

- In Figure 3.8c, the wire is vertical. Because of the wire's position and the direction of the current, the net force acting on the wire drops to zero and the wire's rotation comes to a halt.

For an electric motor, this last situation is unacceptable. In 1832, Hippolyte Pixii recognized this and devised a clever mechanical solution. He attached metal contacts to the armature that reverse the current's direction every time the armature makes half a revolution. This ensures that the force will always be greater than zero and that the armature will continue rotating.

This current reversal is called *commutation*. The metal contacts form a mechanical commutator, which is more commonly referred to as a *brush*. The first practical electric motor, called Jedlik's self-rotor, was a brushed motor, and until the 1960s, brushed motors were the only DC motors available.

 note

A brush is simply a metal contact between a rotor and the external circuit. Brushes are required for commutation, but technically speaking, it's incorrect to say that a brush is a commutator. Nevertheless, most of the literature I've encountered treats "brush" and "commutator" as synonyms. Further, it's safe to assume that every DC motor with a brush also uses a commutator to reverse current.

3.2.2 Types of Brushed Motors

Brushed motors come in three varieties, and the differences between them depend on how the motor generates its magnetic field. If the field is generated by a permanent magnet, the motor is called a permanent magnet DC (PMDC) motor.

The other two types of brushed motors generate magnetic fields using electromagnets. As discussed in Chapter 2, an electromagnet consists of a coil of wire wrapped around an iron core. These coils of wire are called *field windings* or *field coils*. The magnetic field produced by a field winding is proportional to the current flowing through it.

Permanent Magnet DC (PMDC) Motors

PMDC motors are the most popular brushed motors. Because of their permanent magnets, the magnetic field is reliably constant. This means that K_V, the ratio of speed to voltage, is constant. Figure 3.9 illustrates the structure of a brushed DC motor.

One disadvantage of these motors is that permanent magnets lose their magnetization over time. This means the motor gradually produces less torque and speed. This demagnetization accelerates when the armature is driven with large startup currents.

Figure 3.9
Structure of a permanent magnet
brushed DC motor

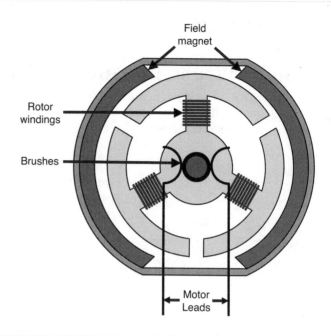

Series-Wound DC (SWDC) Motors

In an SWDC motor, the field winding is connected in series with the rotor winding, which means the current entering the field winding is the same as that entering the armature. Figure 3.10 illustrates the circuit (without electrical losses).

Figure 3.10
A series-wound DC motor

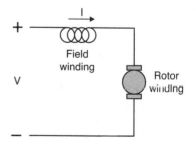

To understand why SWDC motors are useful, it's important to see what happens when the current increases. As explained earlier in this chapter, an increase in current produces an increase in torque. But for an SWDC motor, the increased current also produces an increase in the magnetic field, which further increases the motor's torque. This is why the torque produced by an SWDC motor is much greater than that produced by a PMDC motor.

The disadvantage of using SWDC motors involves speed control. The magnetic field strength changes with current, so the value of K_V changes with current. This makes it hard to reliably set the motor's speed.

Shunt-Wound DC (SHWDC) Motors

In an SHWDC motor, the field winding is placed in parallel with the armature. This means the voltage across the field winding equals the voltage across the armature. The equivalent circuit (without losses) is depicted in Figure 3.11.

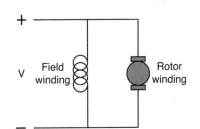

Figure 3.11
A shunt-wound DC motor

This doesn't produce as much torque as a series-wound motor, but the torque-speed curve is generally level. That is, the motor can maintain its speed for different amounts of load. For this reason, shunt-wound motors are commonly used in systems that need to govern the motor's speed reliably.

3.2.3 Advantages and Disadvantages

Brushed motors have improved in performance and reliability since the nineteenth century, but a significant drawback remains: The brush makes contact with the rotor at high speed. As a result, friction erodes the brush over time. It may take months or years, but eventually, every brushed motor will require maintenance to continue functioning.

A second disadvantage is that a brushed motor has to rotate the commutator along with the rotor. This places an additional load on the motor that reduces its efficiency.

Despite these disadvantages, brushed motors are still manufactured and sold in large quantities. The reasons are cost and simplicity. Brushed motors are less complex than brushless motors, so they're less expensive to make. Also, controlling a brushed motor is simpler than controlling a brushless motor, so the circuitry is more cost effective. If you're trying to save money on your project and long-term reliability isn't a major factor, you should consider brushed motors.

3.2.4 Control Circuitry

Controlling a brushed motor is straightforward because the motor's operation is so easy to understand. This section focuses on two types of brushed motor control circuits:

- **Single-direction control**—If the motor only needs to turn in one direction, the circuit can be easily constructed with a transistor.

- **Dual-direction control**—If the motor's direction needs to be changed, an H bridge should be added to the circuit.

In each case, this section presents a basic circuit and describes the components required for its operation.

Single-Direction Control

If a brushed motor only needs to turn in one direction, designing the circuit is easy. The main goal is to enable the controller to turn the motor's current on and off. Earlier in this chapter, I explained how a MOSFET can be used for switching and how the controller can govern the motor with pulse width modulation (PWM).

Figure 3.12 presents a basic circuit that uses a MOSFET for switching. In this circuit, V_{POWER} serves as the main power source and $V_{CONTROLLER}$ is the voltage signal from the controller.

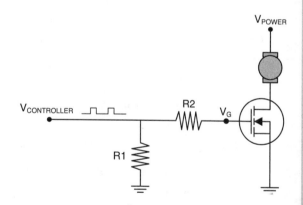

Figure 3.12
Simple circuit for single-direction motor control

This circuit has an important problem. As discussed in Chapter 2, a turning motor generates a voltage called *back-EMF*. If the MOSFET shuts off current after the motor has been running, the motor's back-EMF may damage the transistor. For this reason, control circuits connect a diode in parallel with the motor to provide a path for the back-EMF current. This is called a *flyback diode*, and Figure 3.13 shows what the new circuit looks like.

Some control circuits also insert a potentiometer. This allows a user to directly control the current passing through the motor and thereby increase or reduce the motor's torque and speed.

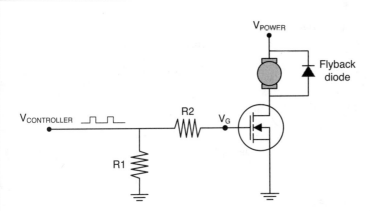

Figure 3.13
An improved circuit for single-direction motor control

Dual-Direction Control

The preceding circuit can turn a motor on and off, but it can't reverse the motor's direction. To make this possible, the circuit needs to be able to reverse the direction of the current flowing through the motor. That is, the circuit needs one path that carries current through the motor in one direction and another path that carries current in the opposite direction. In addition, the circuit needs a way to turn the motor's current on and off.

These requirements can be met by adding an H bridge to the circuit. This component has four electrical switches (S_0–S_3) that can be controlled independently. Figure 3.14a shows what this looks like.

There are many possible states for the switches of an H bridge, but three are particularly important:

- **S_0 and S_3 closed, S_2 and S_1 open**—Current flows through the motor from left to right. This is depicted in Figure 3.14b.

- **S_2 and S_1 closed, S_0 and S_3 open**—Current flows through the motor from right to left. This is depicted in Figure 3.14c.

- **S_0 and S_2 open**—The motor's position is held in place.

It's common to encounter H bridges constructed out of MOSFETs. This is shown in Figure 3.15.

Rather than construct an H Bridge from discrete transistors, it's simpler to use an integrated circuit. The Arduino Motor Shield discussed in Chapter 9, "Motor Control with the Arduino Mega," uses the L298 IC from STMicroelectronics. The RaspiRobot board discussed in Chapter 10, "Motor Control with the Raspberry Pi," relies on the L293DD integrated circuit. If you want to see how H bridges are used in practice, Chapters 9 and 10 are excellent places to look.

If you'd rather not design/construct your own control circuit, you can buy an electric speed control (ESC). These systems receive battery power and produce the pulses needed to power a brushed motor. However, when you select an ESC, make sure it's suitable for your motor and keep in mind that not all ESCs support current reversal.

Figure 3.14
Controlling motor
direction with an H
bridge

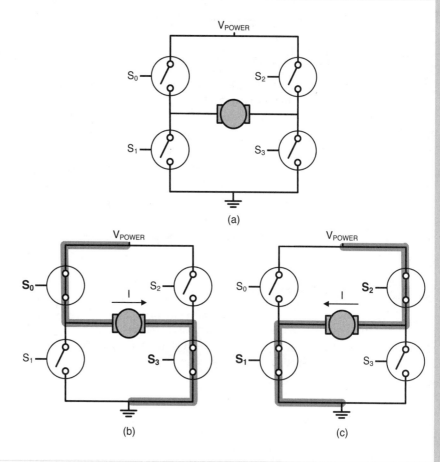

(a)

(b)   (c)

Figure 3.15
H bridge consisting of MOSFETs

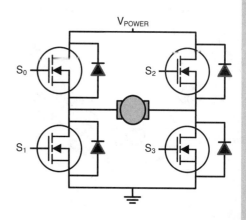

ESCs are convenient for brushed motor systems, but they're essential for systems containing brushless motors. The next section explains why.

3.3 Brushless Motors

The twentieth century brought about a revolution in electronics, as researchers used solid state physics to miniaturize transistors and other circuit elements. With these new integrated circuits, engineers were able to design systems more complex than anything dreamed of before.

As circuit designs grew in complexity, so did the designs of electric motors. In 1962, T.G. Wilson and P.H. Trickey devised a new type of motor that uses electric commutation instead of mechanical commutation. This new motor isn't powered by constant DC electricity, but instead receives timed pulses of DC current. Wilson and Trickey called their new motor a brushless DC motor, or BLDC.

The goal of this section is to present BLDCs in detail. I'll explain how they work and then discuss the two different types of BLDCs: inrunners and outrunners. But first, I want to briefly discuss the scientific principles that make BLDCs possible.

3.3.1 BLDC Structure

BLDCs are more complex and more expensive than brushed motors, but because there's no mechanical contact between the rotor and stator, they're more reliable and efficient. Figure 3.16 presents a cross section of a BLDC's stator and rotor.

> **note**
> The motor depicted in Figure 3.16 is a specific type of BLDC called an inrunner motor. A later section explains the difference between inrunner motors and outrunner motors.

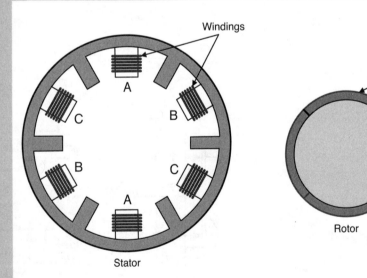

Figure 3.16
A brushless DC motor (BLDC)

As shown in the figure, the structure of a BLDC is completely different from that of a brushed motor. There are multiple current-carrying wires, and they're in the stator instead of the rotor. These wires wrap around iron cores, so they behave as electromagnets. In this chapter and throughout the book, we'll refer to the BLDC's electromagnets as *windings*.

The overall operation of a BLDC isn't hard to understand. The controller delivers positive and negative current to different windings in a sequence, and the rotor spins to follow the change in current. As an analogy, the rotor is like a greyhound in a dog race, constantly following the mechanical rabbit as it runs around the track.

The rest of this section goes into greater depth on three structural aspects of a BLDC: the windings, magnets, and slot/pole configuration.

Windings

In Figure 3.16, the six windings are fixed into positions called *slots*. The controller governs the motor's operation by delivering current to these windings. The process of delivering current to a winding is called *energizing* the winding.

The windings' names are important because windings with the same name are connected. That is, both windings named A receive current at the same time, as do both windings named B and both windings named C. In this manner, the controller only has to deliver three inputs to the motor. For this reason, this BLDC is called a *three-phase motor*. More phases are possible, but most BLDCs are three-phase motors.

A BLDC's windings are energized in a clockwise or counterclockwise manner. As current switches between the windings, the rotor turns at the same speed. Because the motor's speed is synchronized with the changing current, BLDCs can be referred to as *synchronous motors*.

The term *synchronous* is usually associated with AC motors, so it may seem surprising to see it used to describe a DC motor. In fact, the structure of a BLDC is similar to that of a synchronous AC motor in many respects, and I'll explain this further in Chapter 6, "AC Motors."

The main difference between a BLDC and an AC motor is that the current delivered to a BLDC's windings is constant for the duration of the pulse. In an AC motor, the current delivered to the windings is sinusoidal.

At low speeds, a BLDC's rotor may align itself with the stator slots in such a way that it prefers to remain in place than continue rotating. This is called *cogging*, and it can cause the rotor to rotate in a jerky fashion. One solution is to use motors with slotless stators. Slotless BLDCs require more windings due to the greater air gap between the rotor and stator, and are therefore more expensive than regular brushless motors.

Magnets

In a brushed motor, magnets are fixed to the stator. In a brushless motor, magnets are mounted on the surface of the rotor. The rotor in Figure 3.16 has four magnets, but it's common to see BLDCs with many more.

Each magnetic pole in the rotor is called a *pole*. In general, increasing the number of poles increases the torque. BLDC datasheets generally specify how many poles the motor has.

Slot/Pole Configuration

Researchers have spent a great deal of time studying how the number of poles and slots affects a motor's operation. In an *integral slot motor*, the number of slots is a multiple of the number of poles. In a *fractional slot motor*, the number of slots is not a multiple of the number of poles. To reduce cogging, fractional slot motors are generally preferred. The motor depicted in Figure 3.16 is a fractional slot motor because the number of slots, six, isn't a multiple of the number of poles, four.

On occasion, the slots of a brushless motor are referred to as *stator poles* and the magnets are referred to as *magnet poles*. A BLDC's poles may be given as *XN*, *YP*, where *X* is the number of stator poles (slots) and *Y* is the number of magnet poles (magnets).

3.3.2 Inrunner and Outrunner Motors

BLDCs can be divided into two categories depending on the relative positions of the rotor and stator. If the rotor turns inside the stator, as illustrated in Figure 3.16, the motor is an *inrunner*. If the rotor turns outside the stator, it's an *outrunner*. This section discusses both types.

Inrunners

Because the rotor turns inside the stator, an inrunner BLDC motor looks like a regular brushed motor. Judging by appearances, the main difference is that BLDCs have three inputs instead of two. This is shown in Figure 3.17, which depicts the LBA2435 inrunner BLDC from Leopard Hobby.

Figure 3.17
An inrunner brushless motor

Many inrunner motors don't have iron cores. This reduces the iron loss and increases efficiency, but it also significantly reduces the amount of torque the motor can exert. In addition, inrunner motors generally have a low number of poles (commonly only two). For these reasons, inrunner motors usually aren't used to turn propellers without attached gearing. Chapter 7 discusses gears in detail.

To make up for their low torque, most inrunner motors turn at very high speeds. This can be seen by looking at their K_v values, some of which get as high as 7,500 to 10,000 RPM/V.

Outrunners

On an outrunner motor, the rotor is positioned on the outside of the stator. In other words, the permanent magnets spin around the windings. Figure 3.18 presents the structure of an outrunner BLDC.

Figure 3.18
Structure of an
outrunner BLDC

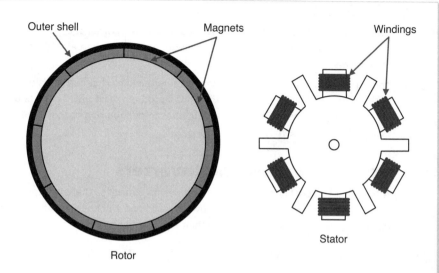

Because the magnets are on the outer shell, outrunners typically have more magnets than inrunner motors. The motor depicted in Figure 3.18 has nine poles, but it's not uncommon to see outrunners with as many as 16.

Outrunner motors don't spin as quickly as inrunners, and their K_v values tend to be around 1,000–2,000 RPM/V. However, they produce significantly more torque. This is why outrunners are commonly employed to spin discs in CD/DVD players. Their high torque has also made them popular in the remote-controlled aircraft community.

Figure 3.19 illustrates the Maytech MTO2830-1300-S outrunner motor. You can see the many internal windings through the front of the motor. This is a common feature that distinguishes outrunners from inrunners.

Figure 3.19
An outrunner brushless
motor

Another difference between inrunners and outrunners involves the motor's shaft. For an inrunner motor, the shaft is connected to the internal rotor. For outrunners, the shaft is connected to the shell.

3.3.3 Controlling BLDCs

Brushless DC motors (BLDCs) are powered with timed pulses of electrical power. The number of inputs is determined by the motor's number of phases. In general, it's safe to assume that a BLDC receives three inputs.

To control a BLDC properly, the pulses must be timed so that current is delivered when the rotor is in the right position. To determine the rotor's position, most BLDC circuits use one of two methods: They measure the back-EMF produced by the motor's rotation or they read the rotor's position using sensors built into the motor.

Control Signals and Inverters

A three-phase BLDC has three inputs that deliver current to the windings. At any time, one input will be set high (V+), one will be set low (V-), and one will be left floating. I'll call these inputs A, B, and C. Figure 3.20 gives an idea how their signals change over time.

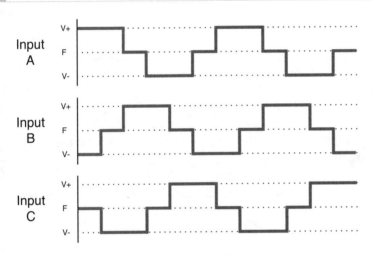

Figure 3.20
Input signals for a three-phase BLDC

For a three-phase BLDC, there are only six unique phase states before they repeat. As the controller energizes the windings through these states, the rotor makes a complete rotation (360°). Therefore, each phase state corresponds to one-sixth of the complete turn, or 60°.

If the controller delivers more current, the motor will exert more torque as it rotates. If the pulses' order and timing is reversed, the motor will turn in the reverse direction. For this reason, BLDC control circuits don't require the H bridges needed to reverse brushed motors.

BLDCs receive power through special switching circuits called voltage source inverters, usually shortened to *inverters*.

 note

Judging from this figure, it may look as though floating inputs are set to 0 V. This is not the case. If an input is floating, it means the controller isn't setting its voltage at all. In other words, the voltage of a floating winding is determined by the motor, not the controller.

Figure 3.21 shows what a MOSFET-based inverter looks like. Each transistor has a flyback diode to discharge current as needed.

Figure 3.21
Switching circuit for three-phase pulse generation

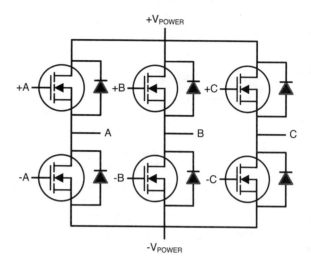

This circuit has six inputs (+A, -A, +B, -B, +C, and -C) that are connected to the controller. Each input is connected to the gate of a MOSFET. Each pair of inputs contributes to an output signal (A, B, or C) that delivers current to the BLDC's windings.

If a positive input is high, the upper MOSFET conducts positive voltage to the corresponding output. If a negative input is high, the lower MOSFET conducts negative voltage to the corresponding output. For example, when the -C input is high, its MOSFET conducts $-V_{POWER}$ to the C output, which is directed to the motor's winding.

Sensored Control

Earlier, I compared the BLDC rotor following the stator's windings to a greyhound chasing a mechanical rabbit around a track. No matter how quickly the greyhound runs, the rabbit must always be kept ahead of it.

The situation is similar for BLDCs. The controller must know the rotor's orientation before it energizes the windings. Some motors have built-in sensors that identify the rotor's orientation and/or speed. If this is the case, the control circuit can use *sensored control*. If the motor doesn't have sensors, the control circuit must rely on *sensorless control*.

A motor's sensors may use optical encoding, magnetic encoding, or variable reluctance, but the most common BLDC sensor is the Hall effect sensor. Many high-end BLDCs, such as those found in larger RC aircraft, have built-in Hall effect sensors.

When current flows through a conductor in the presence of a magnetic field, a voltage is generated that opposes the direction of the current. This is the *Hall effect*. The voltage is called the Hall voltage, V_H, and it's proportional to the product of the current and the magnetic field.

If a BLDC has Hall effect sensors, it will have additional electrical connections that enable the controller to read V_H for each winding. With this information, the controller can determine how the magnetic field is oriented, and therefore how the rotor is oriented. No signal filtering or mathematical computation is necessary.

Sensored motor control is easier to implement and more reliable than sensorless control, but sensored BLDCs require additional circuitry, which means larger motors and higher cost.

Sensorless Control

As discussed in Chapter 2, a motor's rotation produces a voltage called the *back-EMF*. Each winding in a three-phase BLDC produces its own back-EMF, and by measuring the three voltages, a controller can determine how the rotor is oriented.

This raises an important question: If the controller is sensorless, how does it measure back-EMF? The answer is simple. At any time, two of the BLDC's phases are set to a voltage (one positive, one negative) and the third is left floating. The controller measures back-EMF by reading the voltage of the floating winding.

For example, when A is set to V+, B is set to V-, and C is floating, the controller reads the back-EMF of C. When C is set to +V and A is set to -V, the controller reads the back-EMF of B. In this manner, the controller can read the back-EMF of each winding, in turn.

For a BLDC, the back-EMF of each winding has the same approximate shape. As an example, Figure 3.22 depicts the back-EMF voltage across the B winding (solid line) relative to the input current (dashed line).

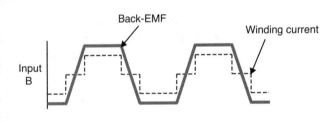

Figure 3.22
Back-EMF for a BLDC winding

Back-EMF is proportional to the motor's speed, so before the controller can measure back-EMF, the motor must already be rotating. Therefore, the controller must start the motor until its windings produce enough back-EMF to allow the controller to determine the rotor's orientation.

This can be confusing, so let me summarize the three steps of sensorless control:

1. The controller delivers current to the motor's windings to start rotation.

2. The controller monitors the back-EMF until it can determine the rotor's orientation.

 note

In a real motor, the shape of the back-EMF isn't as perfect as the figure illustrates. In particular, the back-EMF of a floating winding doesn't change linearly between high and low voltage.

3. After the controller determines the rotor's orientation, it uses this information to synchronize the pulses delivered to the motor's windings.

One popular method of computing the rotor's orientation involves integrating the back-EMF to determine when it equals zero. This is called *zero-crossing detection*, or *ZCD*. When the back-EMF in the floating winding equals zero, the rotor's orientation can be readily obtained. The advantage of the ZCD method is that the computation is simple. However, the back-EMF signal must be filtered to remove noise and the method doesn't work well at high speed. A controller relying on ZCD must limit the motor's speed accordingly.

A more recent method of determining the rotor's orientation makes use of extended Kalman filters (EKF). Kalman filtering requires a great deal of complex math, but unlike ZCD, this method isn't affected by noise. Therefore, if the computation can be performed quickly, an EKF controller allows higher speed than a ZCD controller.

Sensorless control systems are simple to manufacture and cost-effective because no additional hardware is needed. However, a startup period is required so that the controller can measure back-EMF. Further, the process of computing the rotor's orientation places a limit on the motor's maximum speed.

There's one last point I want to mention. As shown in Figure 3.22, the back-EMF of a BLDC winding is approximately trapezoidal in shape. For this reason, BLDCs are frequently referred to as *trapezoidal motors*. In contrast, synchronous AC motors are referred to as *sinusoidal motors* because the back-EMF in their windings is sinusoidal.

3.4 Electronic Speed Control (ESC) Systems

Rather than design and build custom controllers, many designers look for prebuilt circuits called electronic speed control (ESC) systems. Although ESCs are available for brushed motors, because brushless motors are more complex, most ESCs are intended for BLDCs.

To control brushless motors, ESCs come in two types: sensored and sensorless. Most ESCs are designed for sensorless control, and if an ESC's specifications don't specifically mention sensors, it's safe to assume that it's intended to be used with a sensorless motor.

As an example, Figure 3.23 presents an ESC intended to be connected to a sensorless brushless motor.

As in most ESCs, the ESC in the figure has three sets of wires:

- Three wires deliver power to the brushless motor.

- Two wires provide power to the ESC. The red wire should be connected to the positive lead and the black wire should be connected to ground.

- Three wires connect to the controller, possibly through a wireless receiver. These commonly have names such as POS (positive), NEG (negative), and SIG (signal). SIG receives the PWM motor control signals.

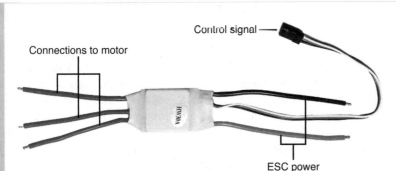

Connections to motor

Control signal

ESC power

Figure 3.23
An electronic speed control
for a sensorless BLDC

When selecting an ESC for your project, it's important to read the specifications. As an example, here are the electrical characteristics of an ESC intended to control a sensorless BLDC:

 note

If an ESC is intended for sensored control, it will have additional connections to receive sensor data from the motor.

- **Input connectors**: Bare wire

- **Maximum current**: 25 A

- **Input voltage**: 7.2–14.14 V Ni-Cd/Ni-MH, 7.4–11.1 V Li-Po

- **Auto cutoff**: Programmable

- **Brake**: Programmable

- **BEC voltage**: Dual BEC circuits

The input voltage identifies the type of battery power that should be connected to the ESC. A later section discusses the important topic of batteries. For now, I want to explain two characteristics of ESCs: battery eliminator circuits (BECs) and programmability.

3.4.1 Battery Eliminator Circuit (BEC)

Many remote-controlled vehicles have separate battery packs for the motors and the wireless receiver. However, if the control circuitry includes a BEC, both systems can draw power from the same set of batteries. Many ESCs contain BECs that provide power to the receiver through the GND and POS wires.

By making a second battery pack unnecessary, a BEC reduces the weight of the RC vehicle and removes the need for checking the charge of two battery packs. The disadvantage is that noise from the motor circuit may interfere with the receiver circuit. Also, if the battery pack runs low on power, it will reduce power to both the motor and the receiver.

3.4.2 Programmability

Some ESCs have operating parameters that can be configured by the user. In many instances, this configuration is made possible through a USB connection to a PC. For the sample ESC, the cutoff and braking features are configurable. The first column of Table 3.1 lists programmable features I've encountered in ESCs. The second column provides a description of each.

Table 3.1 Configurable Parameters of Electric Speed Controllers

Configurable Parameter	Description
Auto-cutoff	Sets the voltage at which the ESC reduces power due to low-voltage state
Brake	Sets the propeller in its braking position when the throttle is in its minimum position
Battery type	Sets the type of battery supplying power to the ESC
Timing	Identifies how quickly pulses are delivered to the BLDC
Reverse	Sets the motor to run in the reverse direction
Reverse delay	The amount of delay time before each reverse in direction
Starting acceleration	Defines how quickly the motor should be accelerated during startup
Current limiter	Sets the maximum amount of current that can be delivered to the motor
Switching frequency	Sets the PWM frequency of the motor control signal

Switching frequency affects the length of time between pulses in the controller's PWM signal. A higher frequency reduces the time between pulses, which makes the motor more responsive to the throttle. When the throttle increases, the PWM duty cycle increases. This delivers more current, which causes the motor to exert more torque as it spins.

3.5 Batteries

For the sample ESC discussed earlier, the battery requirements were given as follows:

Input voltage: 7.2–14.14 V NiMH/Ni-Cd, 7.4–11.1 V Li-Po

In addition, if you look at specifications for motors, you may see statements such as "2–4 Li-Po/5–12 NiMH" or "Max Li-Po Cell: 3s." These seemingly inscrutable descriptions refer to batteries. More specifically, they tell us the type, number, and configuration of batteries that should be used.

Battery-powered motor circuits tend to rely on one of four types of rechargeable batteries. Table 3.2 lists each of them. The second column identifies how much energy the battery can store per kilogram.

Table 3.2 Popular Rechargeable Battery Types

Battery Type	Energy/Mass	Notes
Nickel-cadmium (NiCad or Ni-Cd)	40–60 W-h/kg	Voltage depression, environmentally unfriendly
Nickel-metal-hydride (NiMH)	60–120 W-h/kg	Limited voltage depression
Lithium-polymer (Li-Po)	100–265 W-h/kg	No voltage depression, may explode if overcharged
Lithium-Iron-Phosphate (LiFePO$_4$ or LFP)	90–120 W-h/kg	Chemically stable, constant discharge voltage

Nickel-cadmium (Ni-Cd) batteries used to be popular, and many specifications still include directions for using them in a circuit. However, their usage has waned significantly for two important reasons. The first involves cadmium, which is toxic and requires special disposal procedures. As a result, the European Union has banned the sale of Ni-Cd batteries except for specific applications.

The second disadvantage of Ni-Cd batteries is voltage depression. If a Ni-Cd battery is repeatedly overcharged, its voltage will appear to decrease over time. This effect can be reduced by fully discharging and recharging the battery.

 note

This table doesn't mention lithium-ion (Li-ion) batteries, which are common in smartphones and power tools. This is because Li-Po batteries have the same chemistry as Li-ion batteries, but are packaged differently.

Nickel-metal-hydride (NiMH) batteries are a major improvement on Ni-Cd batteries. A NiMH battery can store more energy than a Ni-Cd battery of similar mass and can discharge a greater amount of power. In addition, it doesn't suffer from voltage depression to the same extent. However, some NiMH batteries lose a significant amount of charge (about 4% each day) if left uncharged.

Lithium-polymer (Li-Po) batteries provide a great deal more energy per kilogram than NiMH batteries. This means that if one circuit has NiMH batteries and another has the same weight in Li-Po batteries, the second circuit will operate longer. Unfortunately, Li-Po batteries can explode if overcharged or overheated. In addition, a Li-Po battery can be damaged if discharged below 80% of its energy capacity. This is why many RC enthusiasts set their Li-Po cutoff voltage at around 3 volts.

The most recent type of battery in Table 3.2 is the lithium-iron-phosphate (LiFePO$_4$ or LFP) battery. These batteries don't provide as much energy per kilogram as Li-Po batteries, but they're more stable and can continue functioning after more charge/discharge cycles.

Having witnessed the explosion of a Li-Po battery, I prefer to use LFP batteries whenever possible. But there are two issues to be aware of. First, they can be hard to find through regular sources. Second, many motor circuits and ESCs are particularly designed for NiMH or Li-Po batteries.

There's one last point to make: If a motor/ESC specification mentions "4s" or "2p," it's referring to how the batteries are connected to one another. Here, "s" refers to *series*, so "4s" refers to four batteries in series. Similarly, "p" refers to *parallel*, so "2p" refers to two batteries in parallel.

3.6 Summary

DC motors are familiar to makers because they're the primary motors in robots, electric vehicles, and 3D printers. Every DC motor needs to switch the direction of input current to ensure that the armature will rotate. This switching is called commutation. Brushed motors use mechanical commutation, and brushless motors use electrical commutation.

Brushed motors are easy to deal with. The armature consists of a single winding that rotates within the stator. The speed of the rotation depends on the input voltage, and the torque depends on the input current. Despite their simplicity, I don't use brushed motors in my circuits because they require maintenance and they aren't as efficient as brushless motors.

Brushless DC motors, or BLDCs, are not easy to deal with. They contain multiple windings, and for most BLDCs, the windings are connected together so that there are three separate inputs. This type of BLDC is called a three-phase motor, and its speed depends both on how quickly the windings are energized and on the incoming voltage. An inrunner is a BLDC whose windings are on the outside and an outrunner has its windings on the inside.

Controlling a brushed motor is simple, but BLDC control can be complicated because the controller needs to know how the rotor is oriented. A sensorless control circuit determines the rotor's orientation by measuring the back-EMF of the motor's floating windings. A sensored control circuit locates the rotor using sensors built into the motor. These may include optical sensors or Hall effect sensors.

Another important consideration is in the choice of batteries. Motor circuits require a great deal of current, and it wasn't so long ago that they relied on Ni-Cd batteries exclusively. But today, motor circuits generally rely on NiMH or Li-Po batteries, which provide a great deal more energy per kilogram. One new option is the $LiFePO_4$ (or LFP) battery, which provides a significant amount of energy and chemical stability.

STEPPER MOTORS

In this and the following chapter, the primary concern is *motion control*—making sure the motor turns with a specific angle and/or speed. This book discusses two types of motors intended for motion control: stepper motors and servomotors. I'll refer to them as *steppers* and *servos*, respectively, and this chapter focuses on steppers.

A stepper's purpose is to rotate through a precise angle and halt. The speed and torque of the rotation are secondary concerns. As long as the stepper rotates through the exact angle and stops, its mission is accomplished. Each turn is called a *step*, and common step angles include 30°, 15°, 7.5°, 5°, 2.5°, and 1.8°.

Due to their simplicity and precision, steppers are popular in electrical devices. Analog clocks, manufacturing robots, and printers (2D and 3D) rely on steppers for motion control. An important advantage is that the controller doesn't have to read the stepper's position to determine its orientation. If the stepper is rated for 2.5°, each control signal will turn the rotor through an angle of 2.5°.

For many applications, we want the step angle to be as small as possible. The smaller the motor's step angle, the greater its *angular resolution*. Another important figure of merit is torque, particularly *holding torque*. A stepper is expected to hold its position when it comes to a halt, and holding torque identifies the maximum torque it can exert to maintain its position.

Modern steppers can be divided into three categories:

- **Permanent motor (PM)**—High torque, poor angular resolution

- **Variable reluctance (VR)**—Excellent angular resolution, low torque

- **Hybrid (HY)**—Combines structure of PM and VR steppers, provides good torque and angular resolution

The first part of this chapter examines these categories in detail. In each case, I'll discuss the motor's fundamental operation and present its advantages and disadvantages. The last part of the chapter explains how steppers can be controlled with electrical circuits.

4.1 Permanent Magnet (PM) Steppers

Small and reliable, permanent magnet (PM) steppers are popular in embedded devices such as disk drives and computer printers. Figure 4.1 depicts the ST-PM35 stepper from Mercury Motor.

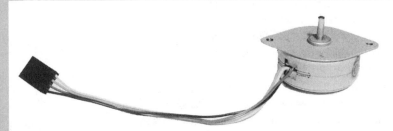

Figure 4.1
A permanent magnet (PM) stepper motor

PM steppers have a lot in common with the brushless DC (BLDC) motors discussed in the preceding chapter. In fact, you can think of a PM stepper as a BLDC whose windings are energized to provide discrete rotation instead of continuous rotation.

4.1.1 Structure

The preceding chapter introduced the brushless DC motor and its two subcategories: inrunners and outrunners. PM steppers are similar to inrunners in many respects, and a good way to introduce them is to compare and contrast them with inrunner BLDCs. Figure 4.2 illustrates the internal structure of a simple PM stepper.

There are five important similarities between PM steppers and inrunner BLDCs:

- Neither motor has a brush or a mechanical commutator (all steppers discussed in this book are brushless).

- The rotor is on the inside, with permanent magnets mounted on its perimeter.

- The stator is on the outside, with electromagnets (called windings) inside slots.

- The controller energizes the windings with pulses of DC current.

- Many of the windings are connected together. Each group of connected windings forms a phase.

Figure 4.2
Internal structure
of a permanent
magnet (PM)
stepper motor

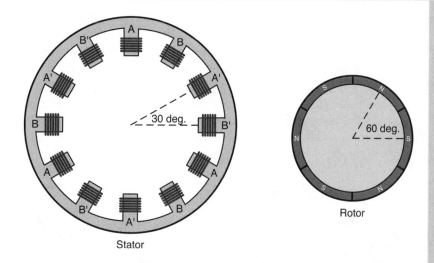

Stator

Rotor

PM steppers are brushless and receive DC pulses from the controller. For this reason, they could be classified as BLDCs. But in this book, as in other literature, we'll only employ the term BLDC for motors that aren't specifically intended for motion control.

Let's look at the differences between the two types of motors. Table 4.1 contrasts the characteristics of PM steppers with those of inrunner BLDCs.

Table 4.1   Contrasting Characteristics of PM Steppers and Inrunner BLDCs

PM Stepper	Inrunner BLDC
Intended for discrete rotation.	Intended for continuous rotation.
Almost always has two phases.	Almost always has three phases.
Controller energizes one or two phases at a time.	Controller energizes two phases at a time and leaves third phase floating.
Many windings and rotor magnets.	Few windings and rotor magnets.

From a structural perspective, the primary difference between PM steppers and inrunners is that PM steppers have more windings and rotor magnets. As it turns out, this is necessary to make the angular resolution as small as possible. The following discussion explains why this is the case.

4.1.2 Operation

To understand how a PM stepper operates, it's crucial to see how its step angle is determined by the number of windings and rotor magnets. This discussion focuses on the motor depicted in Figure 4.2. Its stator has 12 windings and its rotor has six magnets mounted on its perimeter.

PM steppers are generally two-phase motors. In the figure, the different phases are denoted A and B. The windings labeled A' and B' receive the same current as those labeled A and B, but in the opposite direction. That is, if A behaves as a north pole, A' behaves as a south pole.

Each winding has one of three states: positive current, negative current, and zero current. For this discussion, positive current implies a north pole and negative current implies a south pole.

Now let's see how these motors operate. Figure 4.3 illustrates a single turn of a PM stepper. In the windings, a small "N" implies that the winding behaves like a north pole due to positive current. A small "S" implies that the winding behaves like a south pole due to negative current. If a winding doesn't have an N or S, it isn't receiving current.

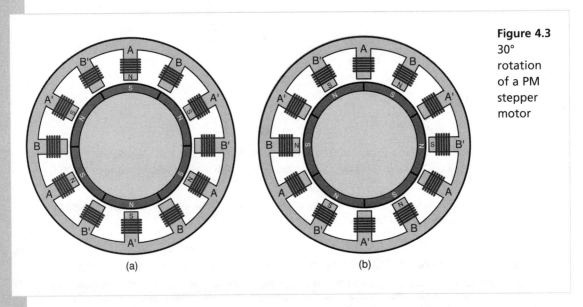

(a)   (b)

Figure 4.3
30° rotation of a PM stepper motor

In Figure 4.3a, A is positive (north pole), A' is negative (south pole), and Phase B isn't energized. The rotor aligns itself so that its south poles are attracted to the A windings and its north poles are attracted to the A' windings.

In Figure 4.3b, B is positive (north pole), B' is negative (south pole), and Phase A isn't energized. The rotor rotates so that its poles align with the B and B' windings. The rotation angle equals the angle between the A and B windings, which means the rotor turns exactly 30° in the clockwise direction. This arrangement of eight windings and six poles is common for PM stepper motors, though others turn at angles of 15° and 7.5°.

In case this isn't clear, let's look at a second movement. Figure 4.4 presents another 30° rotation of a PM stepper motor.

In Figure 4.4a, B is negative (south pole), B' is positive (north pole), and A isn't energized. The rotor is positioned so that its poles align with the B windings.

In Figure 4.4b, A is positive (north pole), A' is negative (south pole), and B isn't energized. The rotor turns exactly 30° in the clockwise direction to align itself between the A windings.

Figure 4.4
Further
rotation
of a PM
stepper
motor

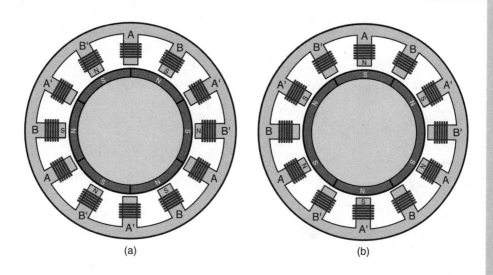

(a)   (b)

The controller's job is to deliver current to the windings so the rotor continues turning in 30° increments. The difference in control signaling is a major difference between steppers and BLDCs. The last part of this chapter discusses the circuitry needed to govern a stepper's operation.

4.2 Variable Reluctance (VR) Steppers

Just as resistance determines the flow of electric current, *reluctance* determines the flow of magnetic flux. In a variable reluctance (VR) stepper, the rotor turns at a specific angle to minimize the reluctance between opposite windings in the stator.

The primary advantage of VR steppers is that they have excellent angular resolution. The primary disadvantage is low torque.

This section presents VR steppers in detail. I'll explain their internal structure first and then show how they rotate as their windings are energized.

4.2.1 Structure

Structurally speaking, variable reluctance (VR) steppers have a lot in common with PM steppers. Both have windings on their stator and opposite windings are connected to the same current source.

However, there are two primary differences between VR steppers and PM steppers:

- **Rotor**—Unlike a PM stepper, the rotor in a VR stepper doesn't have magnets. Instead, the rotor is an iron disk with small protrusions called *teeth*.

- **Phases**—In a PM stepper, the controller energizes windings in two phases. For a VR stepper, the controller energizes every pair of opposite windings independently. In other words, if the stator has N windings, it receives N/2 signals from the controller.

Figure 4.5 illustrates the rotor and stator of a VR stepper. In this motor, the stator has eight windings and the rotor has six teeth.

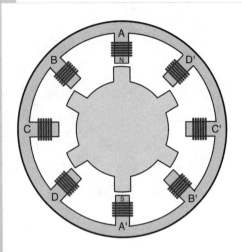

Figure 4.5
Structure of a variable reluctance (VR) stepper

The rotor doesn't have magnets, but because it's made of iron, its teeth are attracted to energized windings. In the figure, the A and A' windings are labeled N and S, which shows how they're energized by the controller. The teeth in the rotor align with these windings to provide a path for magnetic flux between A and A'.

4.2.2 Operation

As illustrated in Figure 4.5, only one pair of teeth is aligned with the windings at any time. When the controller energizes a second pair of windings, the rotor turns so that a different pair of teeth will be aligned. Because the teeth aren't magnetized, it doesn't matter whether a winding behaves as a north pole or as a south pole.

This can be confusing, so Figure 4.6 illustrates the rotation of a VR stepper. In this example, the stepper rotates 15° in a counterclockwise orientation.

In Figure 4.6a, the controller has delivered current to the B and B' windings, and the rotor has aligned itself accordingly. In Figure 4.6b, the C and C' windings are energized. The C and C' windings attract the nearest pair of teeth, which moves the rotor 15° in the clockwise direction.

If you know the number of windings in the stator (N_w) and the number of teeth on the rotor (N_t), the step angle of a VR stepper can be computed with the following equation:

$$Step\ angle = 360° \times \frac{N_w - N_t}{N_w N_t}$$

In Figure 4.6, N_w equals 8 and N_t equals 6. Therefore, the step angle can be computed as 360(2/48) = 15°. The angular resolution can be improved by increasing the number of windings and teeth. With the right structure, the step angle can be made much less than that of a PM stepper.

Figure 4.6
15° rotation
of a VR
stepper

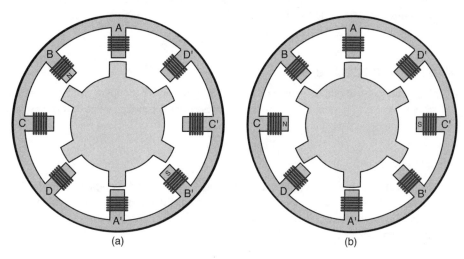

(a)   (b)

However, there's a problem. The torque of a VR stepper is so low that it can't turn a significant load. For this reason, VR steppers are not commonly found in practical systems. In fact, I've only ever seen a handful of VR motors for sale.

To make up for the shortcomings of VR steppers, engineers have designed a motor that combines the resolution of a VR motor and the torque of a PM motor. This is called a hybrid (HY) stepper.

4.3 Hybrid (HY) Steppers

A hybrid (HY) stepper provides the best of both worlds. Like a PM stepper, its rotor has magnets that provide torque. Like a VR stepper, the rotor has teeth that improve the angular resolution. As an example, Figure 4.7 depicts the JK42HW34 hybrid stepper from RioRand.

Figure 4.7
A hybrid (HY) stepper

Hybrid motors have two disadvantages. First, HY steppers can be significantly more expensive than PM steppers. Second, HY steppers are larger and heavier than PM steppers. To see why this is the case, you need to understand their structure.

4.3.1 Structure

If you followed the discussions of PM and VR steppers, HY steppers won't present any difficulty. Their rotors and stators are different from those of either stepper type, but the principle of their operation is similar.

Rotor

If you compare the HY stepper depicted in Figure 4.7 to the PM stepper in Figure 4.1, you'll see that the HY stepper is longer. The reason for this is that the HY stepper rotor has (at least) two rotating mechanisms connected to one another. These are called *rotor poles*, and Figure 4.8 gives an idea of what they look like.

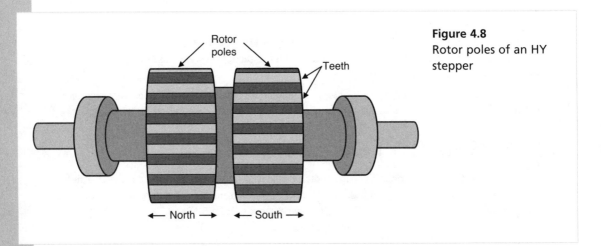

Figure 4.8
Rotor poles of an HY stepper

The rotor poles are magnetized so that one behaves like a north pole and one behaves like a south pole. Each pole has its own teeth, and the teeth of one rotor pole are oriented between those of the other. The angular difference between the two sets of teeth determines the step angle of the motor. The more teeth the stepper has, the better the angular resolution.

The rotor in Figure 4.8 has one pair of rotor poles, but other HY steppers may have two, three, or more pairs. Adding rotor poles increases the stepper's rotational torque and holding torque, but also increases its size and weight.

Stator

The stator windings of a PM stepper or VR stepper are too large to attract/repel the teeth of one rotor pole without repelling or attracting the teeth of the other rotor pole. For this reason, the stator of an HY stepper has teeth that are approximately the same size as the teeth on the rotor. This is shown in Figure 4.9.

Figure 4.9
Toothed stator of an HY stepper

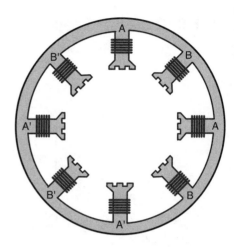

In this figure, each winding has three teeth. In a real stepper, the windings may have many more. If a winding is energized to produce a north pole, its teeth will attract the teeth of the rotor's south pole. If a winding behaves as a south pole, its teeth will attract the teeth of the rotor's north pole.

4.3.2 Operation

Like a VR stepper, an HY stepper can have multiple phases, one for each pair of windings. But the majority of the HY steppers I've encountered are like PM motors. That is, the windings are divided into two phases: A/A' and B/B'. These are the phases labeled in Figure 4.9.

Each phase receives positive current, negative current, and zero current. When one phase is energized, its windings attract the teeth of one rotor pole. When the next phase is energized, its windings attract the teeth of the other rotor pole. Hybrid steppers commonly have 50–60 teeth on a rotor pole, which increases the angular resolution. It's common to see hybrid steppers with step angles as low as 1.8° and 0.9°.

4.4 Stepper Control

Because VR steppers are so scarce, this section focuses on controlling PM and HY steppers, which are almost always two-phase motors. Some PM and HY steppers are bipolar and have four wires. Others are unipolar and have five or six wires.

The terms *bipolar* and *unipolar* identify how the wires are connected to the motor's windings. Before you design a control circuit for a stepper, you should know whether it's unipolar or bipolar as well as the difference between the two types. For this reason, the first part of this section discusses bipolar and unipolar steppers and how to control them.

The last part of this discussion presents different methods of delivering current to a stepper's windings. These methods include half-stepping, which improves angular resolution but reduces torque, and microstepping, which improves angular resolution even further.

4.4.1 Bipolar Stepper Control

A two-phase bipolar stepper has four wires. Figure 4.10 shows how they're connected inside the stepper.

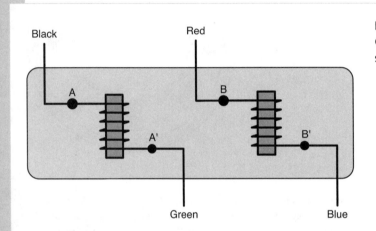

Figure 4.10
Connections of a bipolar stepper

This figure depicts electromagnets and their corresponding phases: A/A' and B/B'. As explained in Chapter 3, "DC Motors," the electromagnet's poles are determined by the nature of the current flow. If current flows from the black wire to the green wire, A will be the north pole and A' will be the south pole. If current flows from green to black, A will be the south pole and A' will be the north pole.

Figure 4.10 identifies the colors of the wires entering the stepper, but these aren't set by any standard. Instead, they follow a convention I've encountered in many bipolar steppers. If you find a stepper whose wires have different colors, the first place to look is the stepper's datasheet. If this doesn't help, you can test the wires with an ohmmeter—the resistance between A and A', like that between B and B', is very small. The resistance between wires in different phases is very high.

To design a circuit that drives a bipolar stepper, you need a means of reversing current in the wires. A common method of accomplishing this involves using H bridges, which were introduced in Chapter 3. An H bridge consists of four switches that, when opened and closed properly, make it possible to deliver current in the forward and reverse directions.

Figure 4.11 shows how an H bridge can be connected to control one phase (A/A') of a bipolar motor. This uses four MOSFETs to serve as the switches.

The current's direction is controlled by setting voltages on the MOSFET gates. When S_0 and S_3 are set high and S_1 and S_2 are low, current travels from A to A', making A the north pole and A' the south pole. When S_1 and S_2 are set high and S_0 and S_3 are low, current travels from A' to A, making A' the north pole and A the south pole. When S_0 and S_2 are left low, the winding is unenergized.

Figure 4.11
Controlling one phase of a bipolar stepper
with an H bridge

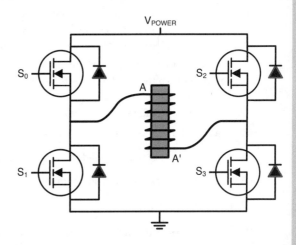

Chapter 9, "Motor Control with the Arduino Mega," and Chapter 10, "Motor Control with the Raspberry Pi," explain how stepper motors can be controlled with real-world circuitry. In both cases, the control circuit contains two H bridges capable of governing both phases of a bipolar stepper motor.

4.4.2 Unipolar Stepper Control

The wiring of a unipolar stepper motor is more complicated than that of a bipolar motor, but the goal is the same: to energize A, A', B, and B' and to set their north/south poles accordingly. To understand how this is done, consider the two circuits depicted in Figure 4.12.

Figure 4.12
Electromagnet circuits with a center tap

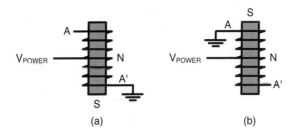

In both figures, V_{POWER} is connected to the center of the electromagnet's winding. This type of connection is called a *center tap*.

In Figure 4.12a, the bottom of the winding is connected to ground. Current flows from the center to ground, energizing the electromagnet and making the bottom of the winding (labeled A') the south pole. The north pole is located at the center.

Now here's the tricky part: The top of the winding isn't connected to anything, so no current flows from the top of the winding to the center. However, the entire iron core is magnetized by the current in the lower wire, which means that the top of the winding also behaves as the electromagnet's north pole. Therefore, in Figure 4.12a, A is north and A' is south.

Figure 4.12b illustrates the reverse situation. The top of the winding is connected to the ground, so current flows from the winding's center to the top. This makes the top of the winding (A) the south pole and the center of the winding the north pole. Because the entire iron core is magnetized, the bottom of the winding (A') also behaves as the north pole.

From a circuit designer's perspective, controlling a two-phase unipolar stepper requires three steps:

1. Provide V_{POWER} to the A/A' and the B/B' windings.

2. For each winding, connect one wire to ground to set the magnetic poles.

3. Leave other wires unconnected.

Figure 4.13 depicts the six wires entering the unipolar stepper: two carry power (V_{POWERA} and V_{POWERB}) and four are connected to A, A', B, or B'. Each of the latter four wires is connected to a MOSFET. When the MOSFET's gate voltage exceeds its threshold, the wire is connected to ground. Otherwise, the wire is left unconnected.

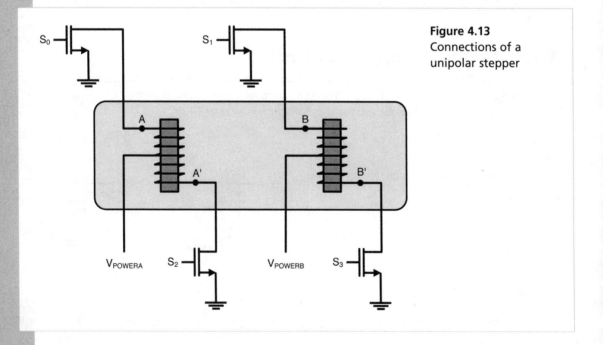

Figure 4.13
Connections of a
unipolar stepper

When a MOSFET switches on, the corresponding end of the winding becomes the south pole. The opposite end of the winding becomes the north pole. For example, when voltage is applied to S_1, the resulting current makes B the south pole and B' the north pole.

 note

This figure doesn't assign colors to any of the wires. This is because I've never found two unipolar steppers that use the same color convention. Check the datasheet to see how the wires should be connected.

Many unipolar steppers have five wires instead of six. For these motors, the two supply wires, V_{POWERA} and V_{POWERB}, are connected together. The other four wires remain unchanged.

Unipolar steppers are easier to control than bipolar steppers because there's no need to manage the switches of two H bridges. However, when a unipolar stepper is energized, only half of the electromagnet is used. Therefore, if a unipolar stepper and a bipolar stepper have the same windings, the unipolar stepper will be half as efficient. This is why I recommend using bipolar steppers whenever possible.

If you ignore the V_{POWER} wires of a unipolar stepper, you can deliver current directly between A and A' and between B and B'. In essence, this is driving a unipolar stepper as a bipolar stepper.

I'd like to make one last point concerning unipolar and bipolar steppers. If you look at a stepper's datasheet, the wiring diagrams won't look like the diagrams presented in this chapter. They represent windings using simpler symbols, and Figure 4.14 shows a sample diagram for a bipolar stepper and a unipolar stepper.

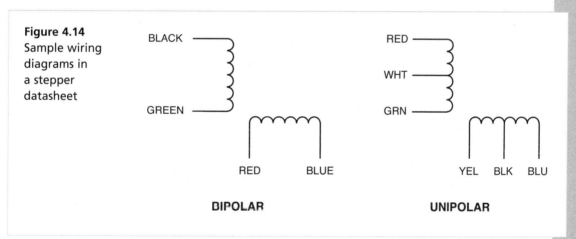

Figure 4.14 Sample wiring diagrams in a stepper datasheet

BLACK

GREEN

RED   BLUE

BIPOLAR

RED

WHT

GRN

YEL   BLK   BLU

UNIPOLAR

Like many datasheets, this figure doesn't identify which winding is A/A' and which is B/B'. This isn't a significant concern. If you replace A/A' with B/B' in a control sequence, the motor's rotation won't be seriously affected.

4.4.3 Drive Modes

This chapter has explained how to operate steppers by energizing one or two winding pairs at a time, but there are a number of different ways to drive a stepper, and this discussion touches on four of them:

- **Full-step (one phase on) mode**—Each control signal energizes one winding.

- **Full-step (two phases on) mode**—Each control signal energizes two windings.

- **Half-step mode**—Each control signal alternates between energizing one and two windings.

- **Microstep mode**—The controller delivers sinusoidal signals to the stepper's windings.

Choosing between these modes requires making tradeoffs involving torque, angular resolution, and power.

Full-Step (One Phase On) Mode

The simplest way to control a stepper is to energize one winding at a time. This is the method discussed at the start of this chapter. Figure 4.15 shows what the signaling sequence looks like when controlling a stepper in this mode.

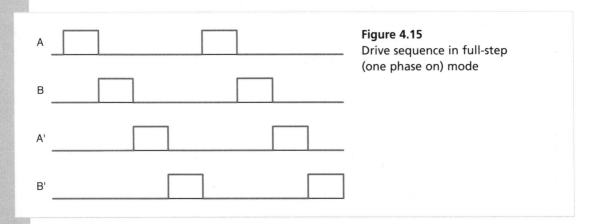

Figure 4.15
Drive sequence in full-step (one phase on) mode

With each control signal, the rotor turns to align itself with the energized winding. The rotor always turns through the stepper's rated step angle. That is, if a PM motor is rated for 7.5°, each control signal causes it to turn 7.5°.

Full-Step (Two Phases On) Mode

In the full-step (two phases on) mode, the controller energizes two windings at once. This turns the rotor through the stepper's rated angle, and the rotor always aligns itself between two windings. Figure 4.16 illustrates one rotation of a stepper motor driven in this mode.

Figure 4.17 shows what the corresponding drive sequence looks like.

The main advantage of this mode over full-step (one phase on) is that it improves the motor's torque. Because two windings are always on, torque increases by approximately 30%–40%. The disadvantage is that the power supply has to provide twice as much current to turn the stepper.

Figure 4.16
Stepper
rotation
in full-step
(two phases
on) mode

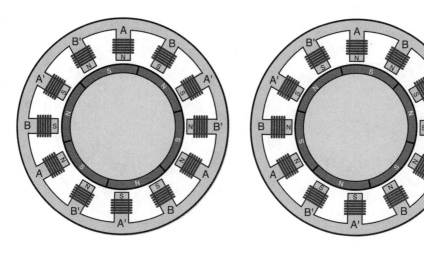

Figure 4.17
Drive sequence in full-step (two phases on)
mode

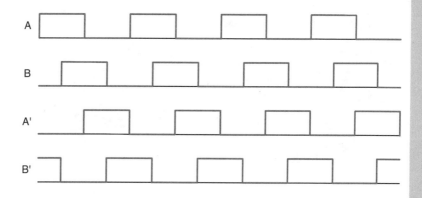

Half-Step Mode

The half-step mode is like a combination of the two full-step modes. That is, the controller alternates between energizing one winding and two windings. Figure 4.18 depicts three rotations of a stepper in half-step mode.

Figure 4.19 illustrates a control signal for a stepper motor driven in half-step mode.

In this mode, the rotor aligns itself with windings (when one winding is energized) and between windings (when two windings are energized). This effectively reduces the motor's step angle by half. That is, if the stepper's step angle is 1.8°, it will turn at 0.9° in half-step mode.

The disadvantage of this mode is that, when a single winding is energized, the rotor turns with approximately 20% less torque. This can be compensated for by increasing the current.

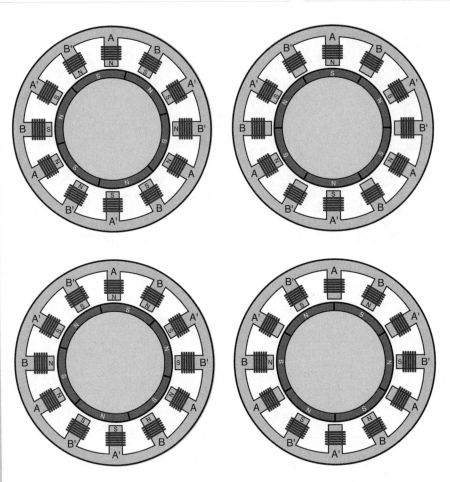

Figure 4.18
Stepper rotations in half-step mode

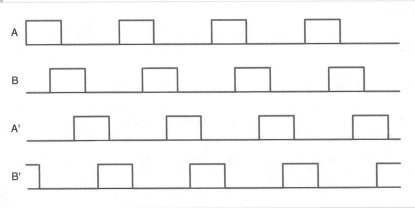

Figure 4.19
Drive sequence in half-step mode

Microstep Mode

The purpose of microstep mode is to have the stepper turn as smoothly as possible. This requires dividing the energizing pulse into potentially hundreds of control signals. Common numbers of division are 8, 64, and 256. If the energizing pulse is divided into 256 signals, a 1.8° stepper will turn at 1.8°/256 = 0.007° per control signal.

In this mode, the controller delivers current in a sinusoidal pattern. Successive windings receive a delayed version of this sinusoid. Figure 4.20 gives an idea of what this looks like.

Figure 4.20
Drive sequence
in microstep
mode

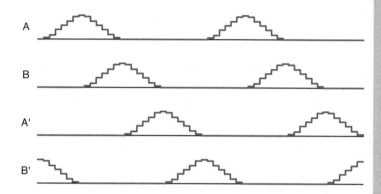

Using this mode reduces torque by nearly 30%, but another disadvantage involves speed. As the width of a control signal decreases, the ability of the motor to respond also decreases. Therefore, if the controller delivers rapid pulses to the stepper in microstep mode, the motor may not turn in a reliable fashion.

4.5 Summary

This chapter has three goals: explain what stepper motors are, present the main types of steppers, and show how steppers can be controlled by a circuit. The first goal is straightforward. A stepper motor is a motor intended to turn at a precise angle (the step angle) and halt. Torque is usually more of a concern than speed, and the torque exerted to hold the rotor's position is called the holding torque.

The first of three types of stepper motor discussed in this chapter is the permanent magnet (PM) stepper. These motors have almost exactly the same structure as the inrunner brushless DC motors discussed in Chapter 3. One significant difference is that PM steppers have many more windings in the stator and magnets in the rotor. These additional windings and magnets make it possible for the PM stepper to turn at step angles such as 15° and 7.5°.

The second stepper type is the variable reluctance (VR) stepper. Like PM steppers, these have windings in the stator. But instead of having magnets on the rotor, the rotor of a VR stepper has teeth. A rotor can support many more teeth than magnets, so the rotor of a VR stepper turns at smaller angles than that of a PM stepper. However, because the teeth aren't magnetized, the rotor is less attracted to the stator's windings. This reduces the stepper's torque to such an extent that VR steppers are rarely encountered in practical systems.

The last stepper type combines the advantages of PM steppers and VR steppers. The rotor of a hybrid (HY) stepper is divided into two or more sections called rotor poles. Each rotor pole is magnetized to behave like a north or south pole, and each has a set of teeth around its perimeter. These teeth are attracted to similar teeth on the stator. Because of the rotor's magnetization, the HY stepper has torque similar to that of the PM stepper. Because of the rotor's teeth, the HY stepper has angular resolution similar to that of the VR stepper. Common step angles of an HY stepper are 1.8° and 0.9°.

When you're designing a control circuit for a stepper, it's important to know whether the motor is bipolar or unipolar. A bipolar stepper has four wires that correspond to the A, B, A', and B' windings. These require H bridges to deliver current in the forward and reverse directions. Unipolar steppers have additional wires that deliver power to the windings. Unipolar steppers are easier to control than bipolar steppers but are less efficient.

The drive mode identifies how the controller energizes the stepper's windings. The simplest drive mode is full-step (one phase on), in which only one winding is energized at a time. For increased torque, the full-step (two phases on) mode energizes two windings at a time. For twice the angular resolution, the half-step mode alternates between energizing one and two windings.

The fourth drive mode is microstep mode. In this mode, the controller divides its control signals into multiple signals of sinusoidal shape. This turns the rotor in tiny step angles to ensure that the rotation is as smooth as possible. Microstepping has been analyzed by many engineers and researchers, but if your system needs smooth motion control, you may want to consider a servomotor instead of a stepper motor. The next chapter presents this fascinating topic.

5

SERVOMOTORS

This chapter focuses on a second type of motor intended for motion control—the servomotor, or *servo*. Whereas steppers rotate through an angle and halt, many servos rotate continuously. Properly controlled, a servo can do everything a stepper can and more. In addition to setting the rotation angle, the controller can configure the servo's rotational speed and acceleration. This added control is the main advantage of servos over steppers.

The main disadvantage is that designing a servo controller is a difficult process. As explained in Chapter 4, "Stepper Motors," a stepper can be controlled with a simple sequence of discrete pulses. With servos, the control signals are more involved. The goal of this chapter is to explain why this is the case and to show how this control can be accomplished.

It's important to note that the term *servomotor* doesn't imply anything about the motor's structure. A servo may be brushed or brushless, AC or DC. The fundamental difference between servomotors and other electric motors is the availability of *position feedback*. A servo sends a signal to the controller that identifies its rotation angle and/or rotation speed. The controller uses this feedback to determine what control signals to send.

Unfortunately, most servos directed toward hobbyists don't provide feedback. I'll refer to these as *hobbyist servos*, and the first part of this chapter presents them in detail. We'll also look at encoders, which are systems that can be connected to motors to provide feedback.

The rest of the chapter focuses on servos that deliver feedback to the controller. Designing a controller for this kind of servo requires mathematical modeling—a model of the servo and a model of the controller's signals.

The benefit of this modeling is that the motor can be made to turn with incredible precision. The drawback is that the mathematical theory has a significant learning curve.

5.1 Hobbyist Servos

Judging from the websites that cater to hobbyists, the main vendors that make hobbyist servos are Hitec, Fitec, Futaba, and Tower Pro. It's possible that one of their offerings provides position feedback, but I've never encountered any.

Most hobbyist servos have a boxlike shape with three wires. Figure 5.1 shows what the FS5106B servo from Fitec (frequently spelled FeeTech) looks like.

Figure 5.1
The FS5106B servomotor from Fitec

The three wires connect the motor to power, ground, and the controller. Additional information can be obtained from its datasheet:

- **Internal structure**: Brushed DC motor

- **Input voltage**: 4.8–6 V

- **Stall torque**: 69.56 oz-in. (4.8 V) or 83.47 oz-in. (6.0 V)

- **No-load speed**: 55.5 RPM (4.8 V) or 62.5 RPM (6.0 V)

- **Running degree**: 180° ± 5°

- **Pulse width range**: 0.7–2.3 ms

- **Neutral position**: 1.5 ms

- **Dead bandwidth**: 0.005 ms

The first four parameters should be clear. The FS5106B servo is based on a brushed DC motor that accepts between 4.8 V and 6.0 V. The torque and speed depend on how much voltage is provided.

The last four parameters may not be clear. They define the nature of the servo's motion and the type of signals needed to control it. This section explores both topics.

5.1.1 Pulse Width Modulation (PWM) Control

Chapter 3, "DC Motors," explained how DC motors can be controlled by sending pulse width modulation (PWM) signals. As a quick review, PWM delivers pulses of varying width to the motor. Figure 5.2 provides a simple example.

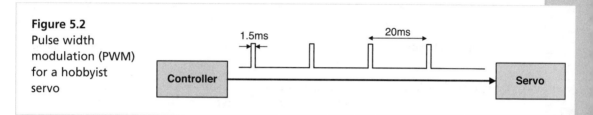

Figure 5.2
Pulse width modulation (PWM) for a hobbyist servo

As given by the "running degree" parameter, the FS5106B's rotor can turn 180°. The controller sets the rotor's angle by sending pulses to the servo and controlling the width of each pulse.

According to the "pulse width range" parameter, the servo responds to pulses whose widths are between 0.7 ms and 2.3 ms. The neutral position is set when the pulse width is at 1.5 ms, which is the pulse width illustrated in Figure 5.2. Similarly, a pulse width of 0.7 ms turns the rotor to the full left position, and a pulse width of 2.3 ms sets the rotor to the full right position.

> 🔍 **note**
>
> Many hobbyist servo specifications don't mention the time between pulses, called the *period*. By convention, this is set to 20 ms, which means the controller should deliver 50 pulses each second.

In theory, we'd like the servo to turn every time the pulse width changes. But in practice, the servo should ignore minor deviations in pulse width that may have been caused by noise. This is the meaning of the "dead bandwidth" parameter, which equals 0.005 ms for the FS5106B. If the pulse width changes by less than 0.005 ms, the servo won't move.

When power is turned off, the servo's rotor stays in its position. This means that, when power is turned on again, the controller may not know the rotor's angle. For this reason, it's common to return the servo to the neutral position as soon as power becomes available.

The best way to understand how controllers govern a servo's behavior with PWM is to look at examples. Chapter 9, "Motor Control with the Arduino Mega," explains how an Arduino circuit board can be programmed to control servomotors. Chapter 10, "Motor Control with the Raspberry Pi," explains how the Raspberry Pi single-board computer can generate PWM pulses for servos.

5.1.2 Analog and Digital Servos

According to its specification, the FS5106B is an analog servo instead of a digital servo. From the controller's perspective, there's no difference between the two—both types are controlled by the same PWM signals. The difference involves the internal circuitry that receives control signals and delivers power to the motor.

When an analog servo receives pulses from the controller, it amplifies them and sends them to the motor to provide power. As the rotor approaches the desired position, the power diminishes to zero. If the widths of the incoming pulses are less than the servo's dead bandwidth (0.005 ms in the case of the FS5106B), the motor won't receive any power.

Digital servos operate in essentially the same way, but a digital servo has a microprocessor that receives pulses from the controller, processes them, and delivers pulses to the motor. The presence of the microprocessor provides three advantages:

- **Lower dead bandwidth**—The processor can respond to pulses that are too small for an analog servo to notice.

- **Higher-frequency power**—The pulses sent by the processor have a higher frequency than those sent by the controller. This makes digital servos more responsive than analog servos.

- **Programmability**—The processor's operating characteristics can be configured by the user.

This last point is particularly interesting. Many vendors of digital servos provide programming tools capable of setting the operating parameters of their motors. For example, a digital servo from Hitec can be programmed with Hitec's HFP-10 programmer, which can set the following parameters:

- **Direction**—Clockwise or counterclockwise

- **Speed**—Rotation speed in RPM

- **Left endpoint/right endpoint**—Maximum angles that can be reached

- **Dead bandwidth**—Minimum pulse width (in microseconds) that the servo will respond to

- **Failsafe**—The servo's behavior when the connection to the controller is cut off

There are two significant disadvantages of using digital servos. The first is cost. Digital servos generally cost twice as much as comparable analog servos. The second disadvantage involves power. Because of its microprocessor, a digital servo requires more power than a comparable analog servo.

A minor disadvantage of digital servos relates to the dead bandwidth. By default, a digital servo responds to small changes in pulse width that an analog servo would ignore. This can be a problem in noisy environments, draining current from the power supply. However, if the servo's dead bandwidth is suitably configured with a programmer, this won't be an issue.

5.1.3 Rotary Encoders

To convert a hobbyist servo (or any electric motor) to a proper servomotor, a mechanism must be attached that identifies the shaft's angle. These feedback elements are called *rotary encoders*. Many types of encoders are available, but for servo applications, there are two main choices:

- **Optical**—A sensor detects light passing through a specially patterned disk.

- **Magnetic**—A sensor detects the moving poles of a magnet.

This discussion presents both types of encoders.

Optical Encoders

Of the many encoders used with servos, optical encoders are the simplest and most common. Their operation is made possible by a disk connected to the motor's shaft. This disk is transparent in some areas and opaque in others.

On one side of the disk, a light source directs light at one portion of the disk. On the other side, an optical sensor measures how much light passes through. The sensor delivers its results to a processor, which may assign a 1 to the presence of light and a 0 to the absence.

The disk's pattern of transparent and opaque regions usually takes one of two forms. Figure 5.3 illustrates both patterns.

Figure 5.3
Optical encoder
disks: incremental
and absolute

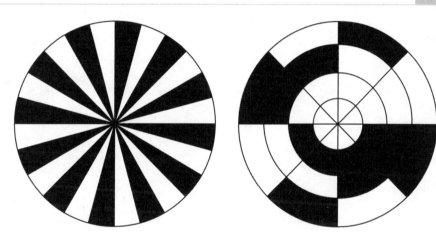

The disk on the left has alternating stripes of transparent and opaque regions. As the shaft turns, the optical sensor measures the time between successive flashes of light. The processor uses this to determine how quickly the motor's shaft is turning. Because it provides speed but not position, this type of encoder is called an *incremental encoder*.

In contrast, the disk on the right of Figure 5.3 is used by *absolute encoders* because it identifies the shaft's angle as well as its speed. In this case, the light shines along an axial stripe that has alternating transparent/opaque regions. The optical sensor detects this light and passes multiple readings to the processor. The microprocessor converts the pattern of light and darkness into a number and uses it to determine the shaft's approximate angle.

Magnetic Encoders

Optical encoders are more common, but magnetic encoders are generally more reliable and provide better resolution. In these encoders, a circular magnet is attached to the shaft. A magnetic sensor is positioned close to the magnet to detect its north and south poles. As the shaft turns, the sensor measures the locations of the changing poles and determines the shaft's angle and speed.

Austria Microsystems produces a family of integrated circuits capable of reading and processing magnet positions. As an example, the AS5145 is a system on a chip that combines Hall effect magnetic sensors and signal processing circuitry. According to its datasheet, this encoder provides angular resolution of 0.0879°. This information is provided through a serial connection and through PWM signals.

5.2 Overview of Servo Control

Chapter 4 discussed the topic of stepper motors and explained how simple they are—a stepper's shaft rotates through a specific angle and stops. But if you measure the shaft's angle as it changes from the initial angle, θ_i, to the final angle, θ_f, you might see something like the graph in Figure 5.4.

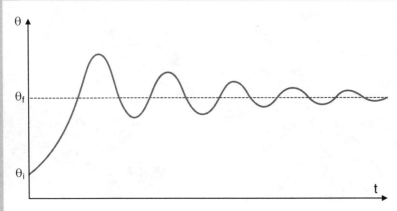

Figure 5.4
Motion of a real-world stepper

As shown, the shaft needs time to ramp up and its angle oscillates before reaching its final angle. Hobbyist servos behave in the same manner. If the motor is intended for a radio-controlled aircraft or a paper printer, this isn't a significant concern.

However, if the intended application requires high-precision motor control, such in the case of robotic surgery, you shouldn't insert a stepper motor and hope for the best. You need a precise understanding of how the motor behaves when it receives signals from the controller. In addition, if the load changes over time, the controller needs to know how to maintain the rotor's motion.

 note

The content in this section makes use of calculus, advanced circuit theory, and other topics that may lie beyond the experience of the average maker. Don't be concerned if it isn't clear. The important thing is to get a sense of the overall method of controlling servomotors using feedback.

The goal of control theory is to make this fine-tuned control possible. To accomplish this goal, it uses mathematical representations of the motor and the controller's signals. This section provides a brief overview of the theory involved.

5.2.1 Open-Loop and Closed-Loop Systems

If a controller receives feedback identifying a servo's shaft angle, it can measure the angle over time to determine the motor's speed and acceleration. If the motor deviates from the desired behavior, the controller will send control signals to reduce the deviation.

This exchange of information—the servo provides its position, the controller provides control signals—forms a loop. For this reason, systems with feedback are referred to as *closed-loop systems*. Systems without feedback are *open-loop systems*. The block diagrams in Figure 5.5 illustrate the difference between these two systems.

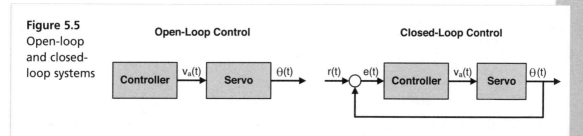

Figure 5.5 Open-loop and closed-loop systems

To analyze closed-loop systems mathematically, we represent each of the signals with functions that change over time. Here are the four functions commonly encountered in servomotor control:

- **θ(t)**—The angle of the servomotor's shaft

- **r(t)**—The desired angle of the servomotor (called the *reference* or the *setpoint*)

- **e(t)**—The deviation (error) between the motor's angle and the desired angle

- **va(t)**—The control signal (voltage) provided by the controller

For the purposes of this chapter, the controller's signal, $v_a(t)$, is an analog function of time. The primary question of servo control is how $v_a(t)$ should be computed from e(t).

As a real-world example, consider a boat whose steering is automated by a computer. Using sonar, visual sensors, and weather databases, the computer determines what course to take. This set of angles forms the system's setpoint, r(t). The controller receives r(t) and delivers a signal, $v_a(t)$, to a servo connected to the ship's rudder. As the servo turns, the rudder changes the boat's heading, denoted as θ(t).

If wind blows the boat off-course, the error, e(t), equals the difference between r(t) and θ(t). The controller receives this error and adjusts $v_a(t)$ to return to the boat to a suitable heading. The goal of control theory is to determine how this new $v_a(t)$ should be computed.

5.2.2 Modeling a Servomotor

The first step in servomotor control is to obtain a mathematical model for the motor's behavior. To be specific, we want to know how its shaft angle changes in response to voltage from the controller.

To derive this relationship, it's important to keep in mind the physical relationships discussed in Chapter 2, "Preliminary Concepts." This chapter presented an equivalent circuit for an electrical motor that included the electrical resistance of the armature, R_a. Figure 5.6 provides a more complete diagram for a motor's equivalent circuit:

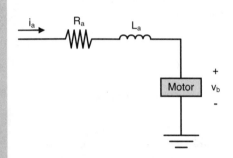

Figure 5.6
Equivalent circuit used in control system design

If i_a is the current passing through the armature's windings, the voltage across the armature's resistance is $R_a i_a$. Similarly, the voltage through the armature's inductance equals L_a multiplied by the time derivative of i_a. Denoting the motor's back-EMF as v_b, the voltage across the armature can be computed with the following equation:

$$v_a = L_a \frac{di_a}{dt} + R_a i_a + v_b$$

Now let's look at the relationship between the motor's torque and angular position. If the shaft's load has a moment of inertia equal to J and a frictional damping coefficient equal to B, the torque, τ, can be related to the shaft angle with the following equation:

$$\tau = J \frac{d^2\theta}{dt^2} + B \frac{d\theta}{dt}$$

This torque can be related to the armature's current with the equation $\tau = K_t i_a$. The shaft angle can be related to the back-EMF with the equation $v_b = K_v (d\theta/dt)$. These relationships were introduced in Chapter 2.

These equations can be combined into a single, lengthy integro-differential equation that relates v_a to θ. However, this equation is hard to understand and nearly impossible to solve directly. To simplify the solution process, control system designers use the Laplace transform.

5.2.3 The Laplace Transform

The Laplace transform makes it easy to deal with complicated equations like those presented in the preceding discussion. This transform converts equations containing derivatives, integrals, and exponentials into equations that can be solved algebraically. A full understanding of its operation requires a deep knowledge of complex analysis, but that's not necessary. To use the Laplace transform, you just have to know three steps:

1. Convert each term of the ugly equation into a simple term based on the variable s. This is accomplished with the forward Laplace transform, denoted **L**{}.

2. Solve the new equation algebraically.

3. Convert each term of the solution back to the t-domain. This is accomplished with the backward Laplace transform, denoted **L**⁻¹{}.

Before proceeding further, let's look at how this magical process works. Consider the following differential equation involving t and the angle θ(t):

$$\frac{d\theta(t)}{dt} - 4\theta(t) = t$$

Solving this normally requires a fair amount of effort, but the Laplace transform makes it easy. The first step is to convert each of the three terms. (Note that the initial condition, θ(0), is a constant value.)

$$L\left\{\frac{d\theta(t)}{dt}\right\} = sL\{\theta(t)\} - \theta(0)$$

$$L\{4\theta(t)\} = 4L\{\theta(t)\}$$

$$L\{t\} = \frac{1}{s^2}$$

Placing these new terms into the original equation produces a new equation in terms of s:

$$sL\{\theta(t)\} - \theta(0) - 4L\{\theta(t)\} = \frac{1}{s^2}$$

With the derivatives gone, the equation is much easier to deal with. Solving for **L**{θ(t)} produces the following equation:

$$L\{\theta(t)\} = \frac{\frac{1}{s^2} + \theta(0)}{s - 4}$$

Using partial fraction expansion, this can be simplified in the following way:

$$L\{\theta(t)\} = -\frac{1}{4s^2} - \frac{1}{16s} + \frac{\theta(0) + \frac{1}{16}}{s - 4}$$

At this point, the second step is finished. The last step is to convert each term of the equation so that it depends on t instead of s. The results are as follows:

$$L^{-1}\{L\{\theta(t)\}\} = \theta(t)$$

$$L^{-1}\left\{-\frac{1}{4s^2}\right\} = -\frac{t}{4}$$

$$L^{-1}\left\{-\frac{1}{16s}\right\} = -\frac{1}{16}$$

$$L^{-1}\left\{\frac{\theta(0) + \frac{1}{16}}{s - 4}\right\} = \left(\theta(0) + \frac{1}{16}\right)e^{4t}$$

Combining these results together gives us the final answer:

$$\theta(t) = \left(\theta(0) + \frac{1}{16}\right)e^{4t} - \frac{t}{4} - \frac{1}{16}$$

Not too hard, is it? The most difficult part is knowing how to convert terms from the t-domain to the s-domain, and vice-versa. Rather than memorize the conversions, engineers rely on tables of general-purpose Laplace transformations. Table 5.1 lists seven of the most common transforms. Other sources provide many more.

Table 5.1 Basic Laplace Transforms

t-Domain	s-Domain
1	$\dfrac{1}{s}$
t^n	$\dfrac{n!}{s^{n+1}}$
e^{at}	$\dfrac{1}{s - a}$
sin(at)	$\dfrac{a}{s^2 + a^2}$

t-Domain	s-Domain
$\cos(at)$	$\dfrac{s}{s^2 + a^2}$
$\dfrac{df(t)}{dt}$	$s\mathbf{L}\{f(t)\} - f(0)$
$\displaystyle\int_0^t f(\gamma)\,d\gamma$	$\dfrac{\mathbf{L}\{f(t)\}}{s}$

A few examples will clarify how these transforms are used in practice:

- If $f(t) = e^{5t} + 1$, then $\mathbf{L}\{f(t)\} = 1/(s - 5) + 1/s$.

- If $f(t) = \sin(2t) + t^3$, then $\mathbf{L}\{f(t)\} = 2/(s^2 + 4) + 6/s^4$.

- If $\mathbf{L}\{f(t)\} = s/(s^2 + 9)$, then $f(t) = \cos(3t)$.

- If $\mathbf{L}\{f(t)\} = s\mathbf{L}\{g(t)\} - g(0)$, then $f(t) = dg(t)/dt$.

The idea of solving an equation by transforming it to another form may seem strange, but the Laplace transform is indispensable in control system design. Most descriptions of control systems don't use the t-domain at all, and rely exclusively on functions in the s-domain.

5.2.4 Block Diagrams and Transfer Functions

The two diagrams in Figure 5.5 are typical of the diagrams used to describe the elements of a control system. Each arrow carries a signal or physical quantity, such as voltage or shaft angle. Each block processes an incoming signal to produce an outgoing signal.

Every block has a corresponding function called its *transfer function*. Unless otherwise stated, this transfer function is multiplied by the input signal to produce the output signal. For example, if a block's function is simply 3, the output signal will equal 3 times the input signal.

Put differently, if a block's input is $X(s)$ and its output is $Y(s)$, the block's transfer function must be $Y(s)/X(s)$. In the case of a servomotor, which converts $V_a(s)$ into $\theta(s)$, the transfer function is $\theta(s)/V_a(s)$.

It's interesting to look at the transfer function of a closed-loop system. Figure 5.7 presents a sample system with two blocks whose transfer functions are $G(s)$ and $H(s)$.

Figure 5.7
Computing the transfer function of a closed-loop system

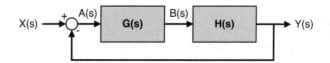

In this diagram, the signal entering the first block is A(s) and the signal leaving it is B(s). Therefore, G(s) = A(s)/B(s). But what's the overall transfer function of the system? That is, given G(s) and H(s), what is Y(s)/X(s)? The answer can be computed in the following way:

$$Y(s) = H(s)B(s) = H(s)G(s)A(s)$$

$$A(s) = X(s) - Y(s)$$

Substituting the second equation in the first equation produces the following results:

$$Y(s) = H(s)G(s)[X(s) - Y(s)]$$

$$Y(s)[1 + H(s)G(s)] = H(s)G(s)X(s)$$

The system's transfer function can be obtained by dividing both sides by X(s) and 1 + H(s)G(s).

$$\frac{Y(s)}{X(s)} = \frac{H(s)G(s)}{1 + H(s)G(s)}$$

If the entire closed-loop system was replaced by a single block, this would be its corresponding transfer function.

5.2.5 Transfer Function of a Servomotor

Having discussed the Laplace transform and transfer functions, the next step is to find the transfer function for a servomotor. This equals $\theta(s)/V_a(s)$, where $\theta(s)$ is the servo's angle and $V_a(s)$ is the control signal.

To obtain the transfer function, the first step is to list the equations that describe how a servomotor works. Earlier in this chapter, I derived one equation involving v_a and another involving τ. In Chapter 2, I presented an equation that related current and torque and another that related voltage and speed equations. There are four equations in total:

1. $v_a = L_a \dfrac{di_a}{dt} + R_a i_a + v_b$

2. $\tau = J \dfrac{d^2\theta}{dt^2} + D \dfrac{d\theta}{dt}$

3. $\tau = K_t i_a$

4. $\dfrac{d\theta}{dt} = K_v v_b$

By applying the Laplace transform, we can convert these equations into a more manageable form. All initial conditions are assumed to equal zero.

1. $V_a(s) = sL_aI_a(s) + R_aI_a(s) + V_b(s)$

2. $\tau(s) = s^2 J\theta(s) + sB\theta(s)$

3. $\tau(s) = K_tI_a(s)$

4. $s\theta(s) = K_vV_b(s)$

Through substitution and algebraic manipulation, we can compute V_a(s) in terms of θ(s). Then it's straightforward to obtain the relationship between the armature voltage and the shaft angle:

$$\frac{\theta(s)}{V_a(s)} = \frac{K_t}{JL_as^3 + (JR_a + BL_a)s^2 + \left(\frac{K_t}{K_v} + R_aB\right)s}$$

This result is important because it gives us precise knowledge of how the servomotor responds to armature voltage. That is, we can multiply this expression by a voltage function (transformed to the s-domain, of course) to determine how the shaft angle will be affected. The following section shows how this transfer function can be used in practice.

5.3 PID Control

After obtaining the transfer function for a servomotor, the next step is to choose a transfer function for the controller block. This block converts the incoming error, E(s), into a voltage signal to the servomotor, V_a(s). Denoting the controller's transfer function as C(s) and the motor's transfer function as M(s), Figure 5.8 presents the closed-loop control system.

Figure 5.8
The closed-loop servomotor system

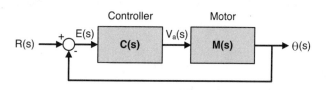

Earlier, I derived the overall transfer function of an closed-loop control system with two blocks, which equaled G(s)H(s)/(1 + G(s)H(s)). Similarly, the overall transfer function for the system in Figure 5.8 is given as follows:

$$\frac{\theta(s)}{R(s)} = \frac{C(s)M(s)}{1 + C(s)M(s)}$$

86 Servomotors

Replacing the servomotor transfer function obtained earlier, this relationship can be expressed with this equation:

$$\frac{\theta(s)}{R(s)} = \frac{C(s)\left(JL_a s^3 + (JR_a + BL_a)s^2 + \left(\frac{K_t}{K_v} + R_a B\right)s\right)}{C(s)\left(JL_a s^3 + (JR_a + BL_a)s^2 + \left(\frac{K_t}{K_v} + R_a B\right)s\right) + 1}$$

The C(s) function identifies the controller's behavior, and there are many possibilities. For servomotors, the most popular method is called *PID control*, where PID stands for *proportional-integral-differential*. In the time-domain, the controller's signal is given as follows:

$$c(t) = K_p e(t) + K_i \int_0^t f(\gamma)d\gamma + K_d \frac{d\,e(t)}{dt}$$

To see why PID controllers are so helpful, we need to examine the three different constant values:

- **Kp, the proportionality constant**—Identifies how the controller should respond to the current value of the error

- **Ki, the integral constant**—Identifies how the controller should respond to the sum of the error over time

- **Kd, the differential constant**—Identifies how the controller should respond to the current slope of the error

Put another way, K_p responds to the present error, K_i responds to the past error, and K_d responds to the predicted future error. By combining these terms, we can configure a PID controller to deal with multiple aspects of the servo's behavior.

Using Table 5.1, the equation for c(t) can be converted to the s-domain. The result is given as follows:

$$C(s) = K_p + \frac{K_i}{s} + sK_d$$

Figure 5.9 shows what the block diagram looks like with a PID controller.

Inserting this expression into the transfer function equation, we get the following result:

$$\frac{\theta(s)}{R(s)} = \frac{\left(K_p + \frac{K_i}{s} + sK_d\right)\left(JL_a s^3 + (JR_a + BL_a)s^2 + \left(\frac{K_t}{K_v} + R_a B\right)s\right)}{\left(K_p + \frac{K_i}{s} + sK_d\right)\left(JL_a s^3 + (JR_a + BL_a)s^2 + \left(\frac{K_t}{K_v} + R_a B\right)s\right) + 1}$$

With the appropriate values for K_p, K_i, and K_d, the controller can set the motor's behavior with remarkable precision. Many research papers have been written on how best to accomplish this, and a full discussion of the different methods lies beyond the scope of this book.

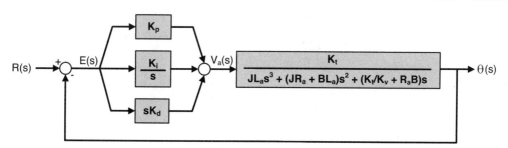

Figure 5.9
Computing the transfer function of a closed-loop system

5.4 Summary

Many sellers of 3D printers provide two versions of their offerings. In the low-cost version, the print-er's motion control is performed by stepper motors, which are inexpensive and precise, but slow because of their discrete rotation. In the premium-cost version, motion control is performed with servos, which rotate continuously with precision and speed. Servos provide many advantages over steppers because of their feedback, but controlling them is significantly more difficult.

With hobbyist servos, feedback isn't a concern. These servos are simply DC motors that can be controlled with PWM signals. They come in two types: digital and analog. A digital hobbyist servo contains a microprocessor that receives the controller's pulses and delivers pulses to the underlying motor. Three advantages of digital servos over analog servos are that they're more responsive, pro-vide greater torque, and can be configured through programming.

Any electric motor can be converted into a servo by attaching a rotary encoder. This provides feed-back to a controller that identifies the shaft's angle. The most common rotary encoders use light sensors that measure light passing through a disk. These optical encoders may be absolute, which means they provide angular position and speed, or incremental, which means they provide speed without identifying the specific angle.

Controlling a servomotor isn't a simple process, and the most difficult aspect involves learning and applying the Laplace transform. Though it can be scary to newcomers, this transform makes it pos-sible to convert ugly equations into simple equations. This conversion is absolutely necessary when dealing with complex electromechanical systems such as DC motors.

The most common method of servomotor control is called PID (proportional-integral-differential) con-trol. A PID controller responds to an error by delivering a signal that sums together three terms. The proportional term equals the error multiplied by a constant, the integral term equals the sum of the current error and past errors, and the differential term equals the slope of the current error. It takes time and effort to properly configure a PID controller, but once the configuration is complete, it can control a servomotor with speed and precision.

6

AC MOTORS

Remote-controlled vehicles and hobbyist devices generally rely on DC motors, but most household/industrial appliances rely on AC motors. The reason is simple—houses and other buildings provide electrical power as alternating current (AC). This is why electric fans and blenders, which have AC motors, can be plugged directly into electrical outlets.

The goal of this chapter is to introduce different types of AC motors and explain their advantages and disadvantages. Because AC motor technology is so old (the first practical AC motors were constructed in the 1880s), a wide range of AC motors is available. They can be categorized in a number of ways, but this chapter classifies motors according to two criteria:

- **Polyphase/single-phase**—The electrical content of the motor's incoming power

- **Synchronous/asynchronous**—The relationship between the motor's speed and the frequency of the incoming power

This chapter introduces polyphase motors first and then presents single-phase motors. The last part of the chapter discusses the fascinating topics of AC motor control and universal motors.

Before I start discussing rotors and stators, I'd like to review the concepts underlying AC power. The better you grasp this topic, the better you'll be able to understand the motors that make use of alternating current.

6.1 Alternating Current (AC)

The fundamental difference between AC motors and DC motors is that the power delivered to an AC motor is sinusoidal. AC power has a number of advantages over DC power, and one major advantage is that AC voltage can be increased and decreased (that is, stepped up and stepped down) with transformers. This makes it possible to transmit AC power over long distances at high voltage and low current. This low current guarantees that the I^2R losses in transmission lines will be as low as possible.

6.1.1 Single-Phase Power

The power provided by residential electrical outlets is single-phase, which means the power is received in a single sinusoid. Figure 6.1 shows what single-phase power looks like.

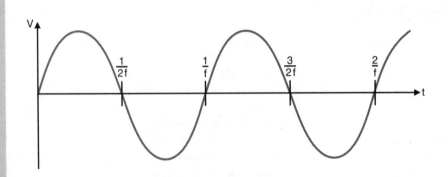

Figure 6.1
Single-phase AC power

In this figure, the sinusoid's frequency is denoted f, which means it completes a cycle in 1/f seconds. In the USA and Canada, f equals 60 Hz and the sinusoid's amplitude is 168 V, which is equivalent to 120 V RMS (root-mean-square). In other nations, it's common to see line power provided at higher voltage (230–250 V RMS) and frequencies of 50 Hz.

6.1.2 Three-Phase Power

Single-phase power is fine for households, but it's insufficient for industrial machines. To meet the greater need, power is delivered with three sinusoids. This is called three-phase power, and Figure 6.2 depicts sinusoidal power provided in three phases, labeled A (solid line), B (dashed line), and C (dotted line).

When you're selecting an AC motor for an application, it's crucial to know what type of power the motor requires. Motors designed for three-phase power won't function properly when given single-phase power, and single-phase motors will likely break when three-phase power is delivered.

Because of their usage in industry, three-phase motors are more common than single-phase motors. But if you're building a product to run on household power, you should take a close look at single-phase motors.

Figure 6.2
Three-phase AC
power

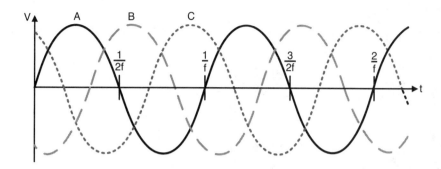

6.2 Overview of Polyphase Motors

Polyphase motors are the workhorses of industry. Cranes, drills, and electric trains all rely on large-scale polyphase motors. These motors come in different types to serve different needs, but their stators all have the same general structure.

The stator of a polyphase motor contains windings (electromagnets) that produce a rotating magnetic field. This rotating field causes the rotor to turn. To understand how polyphase motors work, it helps to understand this rotating field and how it relates to the rotor's speed.

6.2.1 Stators

As discussed in Chapter 1, "Introduction to Electric Motors," an electric motor is the union of two parts: the rotor (which rotates when power is delivered) and the stator (which stays in place). In an AC motor, the structure of the rotor changes according to the motor's type. For example, the rotor of an induction motor is markedly different from the rotor of a permanent magnet synchronous motor.

However, every polyphase AC motor discussed in this chapter has the same stator structure. The stator is always positioned outside the rotor, and its windings receive AC power.

If a motor is intended to receive polyphase power, its windings are grouped into sets called *phases*. The stator has one phase for each phase of input power, and windings in the same phase receive power from the same phase. Figure 6.3 shows what a stator of a three-phase AC motor looks like.

This figure doesn't show the connections between the windings, but A and A' are connected together, B and B' are connected together, and C and C' are connected together. The number of windings per phase is called the number of *poles*, and it's always an even number. The motor displayed in the figure has six windings divided evenly into three phases. Therefore, it has two poles (6/3 = 2).

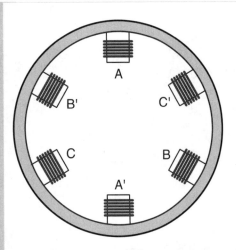

Figure 6.3
Stator of a three-phase AC motor with two poles

6.2.2 Rotating Magnetic Field

Each phase of the input power (A, B, and C) is delivered to the corresponding phase in the stator (A/A', B/B', and C/C'). In Figure 6.3, the three phases are separated from one another in space by 120°. In Figure 6.2, the three voltages are separated in time by an interval that corresponds to 120°. This isn't a coincidence. The alignment of winding position and voltage phase produces a crucial result: a rotating magnetic field in the stator.

The stator's rotating field is vital to the functioning of polyphase motors. To see how it's generated, we can examine the effect of the three-phase voltage on the windings. Figure 6.4 depicts one cycle of power, with markings at times t_0, t_1, t_2, and t_3.

> 🔍 **note**
>
> The following discussion explains how the stator's magnetic field is created. It's not important to understand every detail so long as you're satisfied that the stator of a polyphase motor produces a rotating field.

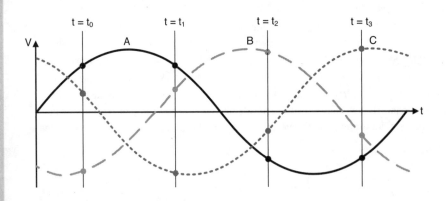

Figure 6.4
Single cycle of three-phase power

The magnetic field produced by a winding is proportional to the current flowing through it, which is proportional to the applied voltage. Therefore, we can gauge the relative strengths of the magnetic fields by comparing their relative voltages.

To determine the windings' magnetic fields at t_0, t_1, t_2, and t_3, we need to find their voltages at these times for each of the three phases. With the maximum voltage set equal to 1, these values are listed in Table 6.1.

Table 6.1 Voltages in Three-Phase Power

	A	B	C
t_0	0.738	−0.952	0.286
t_1	0.738	0.357	−0.976
t_2	−0.762	0.952	−0.310
t_3	−0.738	−0.381	1.0

We can visualize the relative magnetic field produced by each winding by drawing an arrow whose direction is determined by the winding's orientation (0° for A, 120° for B, or 240° for C) and whose length is determined by the winding's voltage. Figure 6.5 depicts these arrows at times t_0, t_1, t_2, and t_3.

Figure 6.5
Magnetic fields at times t_0, t_1, t_2, and t_3

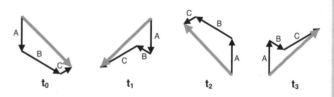

The small arrows represent the field of each winding, and the large gray arrow represents the total magnetic field at the given time. This total field is obtained by arranging the small arrows in sequence. That is, Arrow B starts at the endpoint of Arrow A, and Arrow C starts at the end of Arrow B.

Figure 6.6 presents the same three-phase, two-pole stator as in Figure 6.3, but shows how the magnetic field behaves from t_0 to t_4. The field's direction changes over time, but its strength (represented by the length of the arrow) remains constant.

As depicted in this figure, the magnetic field performs a complete rotation for each cycle of the incoming three-phase power.

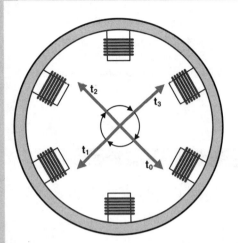

Figure 6.6
Magnetic field in the stator of a three-phase motor

6.2.3 Synchronous Speed

The speed of the stator's rotating field is referred to as the motor's *synchronous speed*. It depends on the frequency of the input power. For a three-phase motor, if the power changes at 60 Hz, the field rotates 60 times per second for a synchronous speed of 3600 RPM.

Many AC motors have more than two poles, and a common number of poles is four. As more poles are added, the synchronous speed decreases because the field has more windings to pass between. If a three-phase AC motor has p poles, the synchronous speed (in RPM) is given by the following formula:

$$n_s = \frac{120f}{p}$$

For example, a three-phase, two-pole motor powered at 60 Hz has a synchronous speed of 3600 RPM, as computed earlier. If a four-pole motor is powered at 60 Hz, then n_s = (120 * 60)/4 = 1800 RPM.

6.2.4 Power Factor

For a DC motor, computing a motor's input power is simple. Input power = VA, where V is the input voltage and A is the input current. The computation is easy because the voltage and current are always proportional to one another.

For AC motors, determining the input power isn't as simple. This is because the input current and voltage change with the same frequency, but the crests and troughs of the two sinusoids usually don't align. In this case, they are said to be *out of phase*. This is shown graphically in Figure 6.7.

Figure 6.7
Out-of-phase voltage
and current

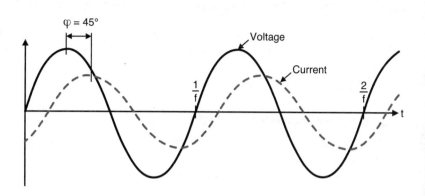

In this figure, the voltage and current have the same frequency, but their crests are separated by a time equal to one-eighth of the cycle, or 1/8f. A full cycle corresponds to an angle of 360°, so the angular interval between the two crests is 360°/8 = 45°. This angle, denoted φ, is called the *phase angle*. It's positive if the voltage precedes the current and negative if the current precedes the voltage. Ideally, the phase angle is 0°.

When you're examining an AC motor, an important performance parameter is its *power factor*, which equals the cosine of φ. This tells you how much of the total input power is being used for work. This useful power is called *real power*. Ideally, the power factor equals 1 because cos(0) = 1. But in practice, this value is reduced by a motor's capacitance or inductance.

The power factor, commonly denoted PF, can be expressed in another way:

$$PF = \frac{power\ able\ to\ do\ work\ (real\ power)}{total\ power\ supplied\ to\ motor}$$

In this fraction, the denominator can be calculated by multiplying the amplitude of the current by the amplitude of the voltage.

An example will make this clear. If a motor has a power factor of 0.8, 80% of the input power will be used to perform work. If this motor draws 4 A at 50 V, the amount of real power will be (4 A)(50 V) (0.8) = 160 W.

If the motor has three phases, the equation for the power factor changes slightly:

$$PF = \frac{power\ able\ to\ do\ work\ (real\ power)}{\sqrt{3} \cdot total\ power\ supplied\ to\ motor}$$

For example, suppose each phase of a three-phase motor draws a current of 6 A at a voltage of 200 V. If the power factor is 0.75, the real power is computed as follows:

$$Real\ Power = \sqrt{3}(0.75)(6A)(200V) = 1559\ W$$

6.3 Asynchronous Polyphase Motors

If a motor is *asynchronous*, the speed of its shaft doesn't equal the motor's synchronous speed. To be more specific, the rotor of an asynchronous motor turns at a speed less than the synchronous speed. To see why this is the case, it's crucial to understand the principle of electromagnetic induction.

6.3.1 Electromagnetic Induction

If a conductor is brought into a region with a changing magnetic field, it receives a difference in voltage across its surface. This phenomenon is referred to as *electromagnetic induction*, and the voltage is called the *induced voltage*. The magnitude of the voltage is proportional to how quickly the magnetic field is changing.

Induced voltage produces a current in the conductor. When a current-carrying conductor is placed inside a magnetic field, it receives a force that causes the conductor to move. This is the fundamental principle behind all asynchronous motors, which are commonly called *induction motors* or *AC induction motors (ACIMs)*.

In case this isn't clear, let me express the basic operation of an induction motor in three steps:

1. As discussed in the preceding section, the stator of an AC motor creates a changing magnetic field in response to polyphase power.

2. The rotor of an induction motor has conductors instead of magnets. When these conductors enter the stator, each receives an induced voltage.

3. The induced voltage produces a current in each conductor. As a result, a force is exerted on each conductor, and this force turns the rotor.

The force produced by induction is present only when the rotor's conductors intersect the stator's magnetic field. If the rotor's conductors travel at the same speed as the field, no force will be produced because there's no intersection. Therefore, in an induction motor, the rotor always turns at a speed lower than that of the stator's field.

The relationship between an asynchronous motor's speed and its synchronous speed is called *slip*. Denoting the motor's speed as n, its slip can be computed with the following equation:

$$s = \frac{n_s - n}{n_s}$$

When an asynchronous motor receives power, there's a delay before the rotor starts turning. Therefore, when a motor is started, n = 0 and s = 1. The slip reaches its minimum value when no load is attached to the shaft. As the load increases, n decreases and the slip increases.

Slip is usually expressed as a percentage of the synchronous speed. Therefore, if the synchronous speed is 3,600 RPM and the rotor turns at 3,200 RPM, the slip equals (3600 − 3200)/3600 = 0.111 = 11.1%.

6.3.2 Current and Torque

Like many motors, an asynchronous motor requires a significant amount of current to start its operation. In some cases, this startup current can be four to eight times as large as the current used during regular operation. This increased current means increased torque, and in some asynchronous motors, the starting torque can be two or three times larger than the torque exerted in full-loading conditions.

When a load is attached to the shaft, the rotor's speed decreases and its conductors spend more time intersecting the stator's rotating field. This induces greater current in the conductors, which produces a greater torque on the rotor. Therefore, the motor's torque increases with the applied load. This is shown in the graph in Figure 6.8.

Figure 6.8
Typical speed-torque graph for an asynchronous motor

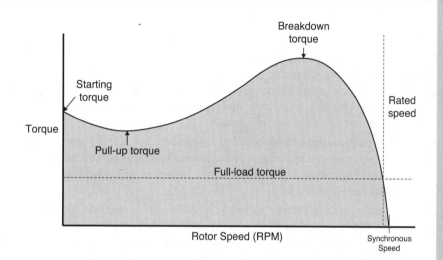

As the load increases, the torque increases until it reaches the *breakdown torque*. The *pull-up torque* is the minimum torque between the starting torque (also called the *locked rotor torque*) and the breakdown torque.

The speed given on a motor's datasheet is its *rated speed*. When the motor runs at its rated speed, the torque it exerts is its *full-load torque*. These properties are illustrated in Figure 6.8 with dashed lines. The motor's slip can be computed using its rated speed and synchronous speed.

The motor exerts no torque when its speed matches its synchronous speed. This is because the rotor's conductors aren't intersecting the rotating magnetic field.

6.3.3 Squirrel-Cage Rotor

The oldest and most popular type of asynchronous motor has a cylindrical rotor with conductors embedded into its surface in a striped pattern. This rotor looks like a running wheel used by gerbils and hamsters, so it's called a *squirrel-cage rotor*. Figure 6.9 gives an idea of what it looks like.

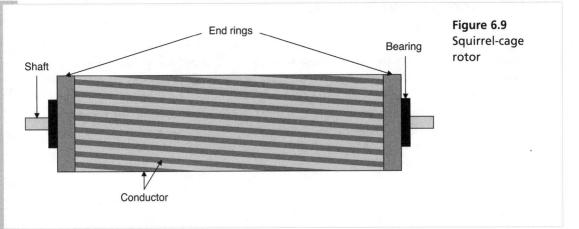

End rings

Bearing

Shaft

Conductor

Figure 6.9
Squirrel-cage
rotor

The conductors along the surface of the rotor are usually made of copper or aluminum, and the core is commonly composed of steel. The conductors are connected to one another through the metal end rings. To reduce cogging, the conductors are skewed at an angle. As discussed in Chapter 3, "DC Motors," cogging occurs when the rotor temporarily locks in place as it turns. This causes the motor to operate in a sporadic, jerky fashion.

The chief advantage of the squirrel-cage rotor is its simplicity and reliability. The rotor has no commutators, electromagnets, or moving parts, so it can be easily fixed or replaced. Also, because permanent magnets are so expensive, squirrel-cage rotors are much cheaper than motors whose rotors have permanent magnets.

Because of the simplicity, reliability, and low cost, AC motors with squirrel-cage rotors can be constructed at large sizes. Whereas DC motors can fit in the palm of your hand, motors with squirrel-cage rotors can occupy an entire room. Most of the motors in pumps, blowers, heaters, and air conditioners are AC motors with squirrel-cage rotors.

6.3.4 Wound Rotor

The main drawback of using a motor with a squirrel-cage rotor is that its speed-torque characteristic can't be changed. The rotor's conductors can't be accessed from outside the motor, so the only way to increase or decrease the motor's speed is to change the frequency of the incoming power.

To improve on this, engineers designed a motor whose rotor contains coils of conductive wire instead of conductive bars. This type of motor is called a *wound-rotor motor* or a *slip-ring motor*. For the sake of brevity, I'll refer to these motors as WRIMs (wound-rotor induction motors).

Because of its coils, a WRIM's rotor is similar to that of a brushed DC motor. However, there are two important differences:

- Instead of a brush, the coils of a WRIM connect to external circuitry through slip rings on one end of the rotor. There is one slip ring for each phase.

- Unlike the coils in a brushed DC motor, the WRIM's coils don't receive power from outside the motor.

This second point is important to understand. As in all induction motors, a WRIM's rotor receives current through induction, not through external power. Instead of power, the rotor's coils are connected to variable resistances through slip rings on the shaft. Figure 6.10 gives an idea of what the equivalent circuit looks like.

Figure 6.10
Rotor circuit of a wound-rotor induction motor

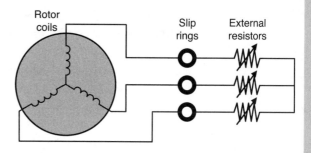

Increasing the resistance of the WRIM's coils reduces the strength of the stator's magnetic field. This significantly reduces the amount of starting current required by the motor.

For large machines, this can save a great deal of power and can reduce the chances of startup failure. The graph in Figure 6.11 illustrates the approximate relationship between an induction motor's current and speed for different levels of resistance.

Figure 6.11
Effect of resistance on a WRIM's current draw

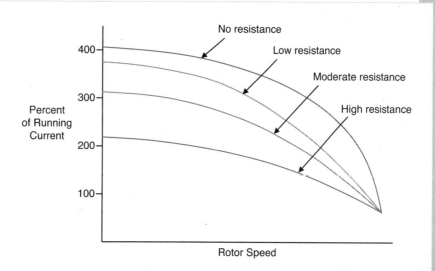

Because of the reduced startup current, the WRIM's initial speed is lower than that of a comparable motor with a squirrel-cage rotor. This decreased speed means that the conductors intersect the stator's magnetic field more often. This interaction increases the rotor's torque, which means a WRIM's starting torque is usually greater than that of a squirrel-cage motor. Figure 6.12 gives an idea of how a typical WRIM's torque changes with resistance.

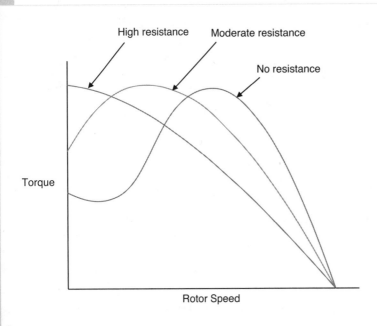

Figure 6.12
Effect of resistance on a WRIM's torque

As shown in the figure, increasing the rotor's resistance moves the motor's breakdown torque to the left. If the breakdown torque is brought all the way to the left, the motor's starting torque will be significantly larger than the torque without resistance.

The drawback to the WRIM's resistance is that its overall torque decreases as the load increases. For this reason, it's common for the motor circuit to reduce or remove the WRIM's resistance as the load increases.

6.4 Synchronous Polyphase Motors

If a motor's rotational speed is equal to its synchronous speed, it's a synchronous motor. These have the same stator structure as asynchronous motors, but the rotors are significantly different. Based on rotor structure, synchronous motors can be divided into three categories:

- **Doubly excited synchronous motors**—The rotor has windings that receive electrical power.

- **Permanent magnet synchronous motors**—The rotor has permanent magnets embedded into its perimeter.

- **Synchronous reluctance motors**—The rotor has teeth that turn with the stator's magnetic field.

This section discusses the motors in each of these categories and explains how their rotors make synchronous operation possible.

6.4.1 Doubly Excited Synchronous Motors

One type of synchronous motor has windings (electromagnets) in its rotor that receive current from outside the motor. These motors are called *doubly excited* because the rotor and stator both receive power. But the rotor receives DC current instead of AC current, and the current is delivered through slip rings on the rotor's shaft.

The rotor of a doubly excited motor has windings and conductors on its perimeter. In this way, it combines aspects of a squirrel-cage rotor and a wound rotor. Figure 6.13 gives an idea of what this looks like.

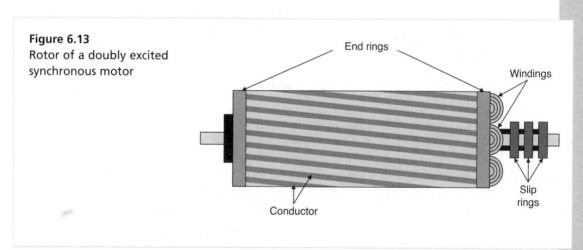

Figure 6.13
Rotor of a doubly excited synchronous motor

As shown, the rotor's outer shell has conductors similar to those found on a squirrel-cage rotor. They serve an important purpose. Doubly excited motors, like many synchronous motors, can't start by themselves. That is, the DC power directed to the rotor's windings isn't sufficient to start the rotor turning. However, when voltage is induced in the rotor's conductors, the force starts the motor.

After the motor starts and current is provided through the slip rings, the rotor's windings behave as electromagnets. The electromagnet's north and south poles are attracted to the opposite poles in the stator's rotating field. This attraction turns the rotor, which moves at the same speed as the stator's field.

6.4.2 Permanent Magnet Synchronous Motors

A permanent magnet synchronous motor, commonly called a PMSM, has permanent magnets mounted on the rotor. In this way, it has essentially the same structure as the brushless DC motor (BLDC) discussed in Chapter 2, "Preliminary Concepts." Figure 6.14 depicts the rotor and stator of a simple PMSM.

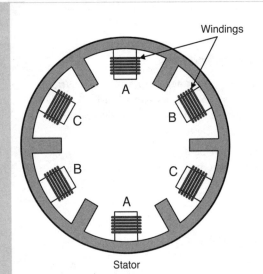

Windings

A

C

B

B

C

A

Stator

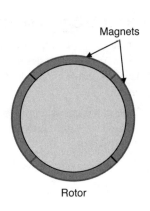

Magnets

Rotor

Figure 6.14
Cross-section of a PMSM's rotor and stator

The fundamental difference between the PMSM and the BLDC is how the stator's windings are energized. In a BLDC, the stator's windings receive pulses of DC current. In a PMSM, the stator's windings receive AC power similar to that depicted in Figure 6.2.

Because of the difference in power, the back-EMF of a PMSM has a different shape than that of the BLDC. Whereas the back-EMF of a BLDC is approximately trapezoidal in shape, the back-EMF of a PMSM is sinusoidal in shape. For this reason, BLDCs are frequently referred to as trapezoidal motors and PMSMs are referred to as sinusoidal motors.

6.4.3 Synchronous Reluctance Motors

The simplest synchronous motors are the reluctance motors. Their rotors don't receive power and they don't have permanent magnets. Instead, the rotors of a reluctance motor are made of a ferromagnetic material (usually iron) that responds to magnetic fields.

On the rotor's perimeter, regions are removed to concentrate the stator's field into the remaining regions. These remaining regions are called *salient poles*, and Figure 6.15 shows what they look like.

The principle behind the motor's operation is just like that of the variable reluctance steppers discussed in Chapter 4, "Stepper Motors." The rotor follows the magnetic field to minimize reluctance (the magnetic equivalent of resistance) in the stator. The main difference between the variable reluctance stepper and the synchronous reluctance motor is the power supplied to the stator's windings: The stepper receives DC pulses, and the synchronous motor receives AC power.

Synchronous reluctance motors operate at low power, and they're inexpensive because they don't require magnets or coils. The main disadvantage is they exert much less torque than other motors. For this reason, they can be hard to find. Only a few companies, such as ABB, manufacture these motors in significant quantities.

Figure 6.15
The synchronous reluctance motor

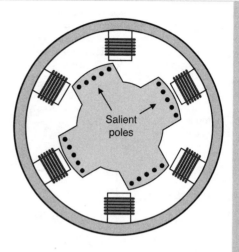

Salient poles

6.5 Single-Phase Motors

If you intend to build a motorized device that runs on household power, you should focus on single-phase motors. These follow the same essential principles as polyphase motors, but there's one major difficulty: Single-phase power can't produce a rotating magnetic field in the stator in the way that three-phase power can. Without the rotating magnetic field, the motor can't start on its own.

For this reason, engineers have devised many methods to generate rotating fields using single-phase power. Each method corresponds to a different type of motor, and this section discusses three types of single-phase motors:

- **Split-phase motor**—Uses a main and auxiliary winding

- **Capacitor-start motor**—Inserts a capacitor into the stator circuit

- **Shaded-pole motor**—Blocks off part of a winding to serve as a second pole

These are all induction motors with squirrel-cage rotors. As discussed earlier, the rotors turn because of the current induced in their conductors. The speed of rotation is slightly less than the synchronous speed, which is determined by the changing magnetic field in the stator.

I've heard about single-phase synchronous motors, but I've never actually seen one. All the single-phase motors I've encountered belong to the three categories just mentioned.

6.5.1 Split-Phase Motors

The term *phase-splitting* refers to obtaining two signals with different phases from one single-phase signal. A split-phase motor generates a rotating magnetic field in the stator by phase-splitting the single-phase input power.

To accomplish this goal, a split-phase motor has two windings in its stator: a main winding and an auxiliary winding. They're connected to the single-phase power in parallel at an angle of 90°. This is shown in Figure 6.16.

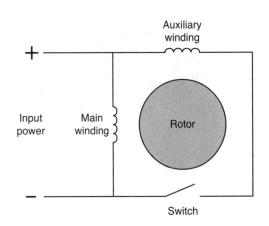

Figure 6.16
Stator circuit of a split-phase motor

Compared to the main winding, the auxiliary winding has a high resistance and a low inductance. This ensures that the currents in the two windings are out of phase with one another. The ideal phase difference is 90°, but the actual phase difference is usually between 30° and 40°. Thankfully, this is sufficient to produce a rotating magnetic field.

The small phase difference produces a small magnetic field, which results in low torque. But because of the phase-splitting, split-phase motors are capable of starting by themselves. After the motor starts and the rotor's speed approaches the rated speed, a switch opens and removes power from the auxiliary winding.

6.5.2 Capacitor-Start Motors

A *capacitor-start* motor has the same essential structure as a split-phase motor, but improves on it by adding a capacitor in series with the auxiliary winding. Figure 6.17 shows what the stator circuit looks like.

The capacitor increases the phase difference between the current in the main winding and the current in the auxiliary winding. The increase in phase difference produces a corresponding increase in starting torque, which is the primary advantage of capacitor-start motors over split-phase motors. In addition, the capacitor reduces the current needed to start the motor's operation.

In a capacitor-start motor, the circuit's switch opens when the motor's speed approaches its rated speed. This disconnects the capacitor and the auxiliary winding from the power.

One variation on the capacitor-start motor is the *permanent split capacitor* (PSC) motor. This removes the switch from the circuit, so the auxiliary winding and its capacitor are always connected to the main winding. The removal of the switch increases the motor's reliability and the capacitor improves the motor's power factor, but the starting torque isn't as large as that of a capacitor-start motor.

A second variation is the *capacitor-run motor*. This adds a second, larger capacitor in parallel to the first. Figure 6.18 shows what the stator circuit looks like.

Figure 6.17
Stator circuit of a capacitor-start
motor

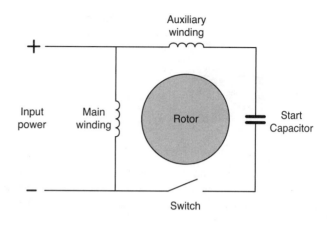

Figure 6.18
Stator circuit
of a capacitor-
run motor

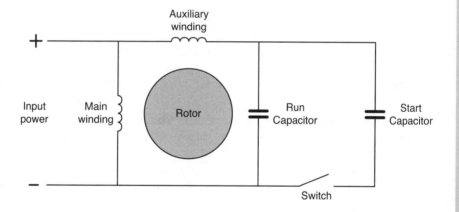

Capacitor run motors provide the same high starting torque as capacitor-start motors and the same high power factor as permanent split capacitor motors. However, the complexity of the circuit increases the cost as well as the potential for electrical and mechanical problems.

6.5.3 Shaded-Pole Motors

The shaded-pole motor is the oldest of the single-phase motors discussed here and is also one of the least expensive. This type of motor doesn't have any auxiliary windings, and it doesn't add any new components. Instead, it changes the shape of the stator's main winding.

To be specific, a portion of the iron core is cut away and the remaining portion is encircled by a conductive ring. Figure 6.19 shows what this looks like.

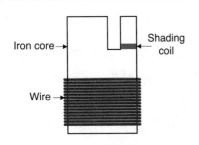

Iron core →

Wire →

Shading
coil

Figure 6.19
Main winding of a shaded-pole motor

The conductive ring is called a *shading coil* and it serves an important purpose. Like the conductors in the rotor, the ring receives an induced voltage that produces an induced current. This current is out of phase from the rest of the current in the winding. As a result, the two out-of-phase currents produce a rotating magnetic field.

The generated field is so small that shaded-pole motors should only be used in low-torque applications. Their main advantage is that they're simple and inexpensive to construct.

6.6 AC Motor Control

Most AC motors don't have control circuitry—many operators plug the motor into the power outlet and control it with the On/Off switch. An AC motor's input power is determined by its locale (120 V/60 Hz in the U.S. and Canada), so these motors generally run at a fixed speed with a fixed torque.

One type of motor makes it possible to control speed and torque by connecting or disconnecting poles in the stator. These motors, called *multispeed motors*, are used in adjustable devices such as ceiling fans.

To provide additional control, engineers have constructed two devices: eddy-current drives and variable-frequency drives (VFDs). The first part of this section discusses eddy-current drives, and the rest of this section discusses the operation of VFDs.

6.6.1 Eddy-Current Drives

An eddy-current drive isn't a traditional motor controller. That is, it doesn't change any aspect of the motor's operation. Instead, it converts the rotation of the motor's shaft to the rotation of a second shaft. This second shaft turns with the desired torque and speed.

An eddy current drive consists of four parts:

- A fixed-speed induction motor.

- The clutch, which connects to the motor's shaft and turns the output shaft.

- A tachometer, which measures the position of the output shaft.

- The controller, which reads data from the tachometer and delivers current to the clutch.

Figure 6.20 shows how these components are connected.

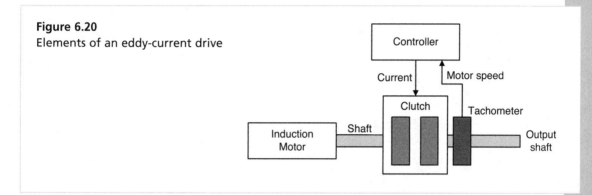

Figure 6.20
Elements of an eddy-current drive

As the motor operates, the clutch converts the rotation of its shaft into the rotation of the output shaft. An adjustable magnetic field inside the clutch determines how the output torque is related to the input torque. The speed of the output shaft is determined by the controller, which reads the position of the output shaft from the tachometer and delivers current to the clutch.

6.6.2 Variable-Frequency Drives

A variable-frequency drive (VFD) is connected between a motor and line power, and its purpose is to generate power for the motor with the desired voltage and frequency. By using a VFD, you can tailor the input power to meet the motor's needs. This can save a significant amount of money and extend the motor's life span.

The operation of a VFD consists of two steps:

- It converts input AC power to DC power. This entails rectification and smoothing.

- An inverter generates a pulse width modulated (PWM) signal with the desired voltage and frequency.

It's important to understand that VFDs deliver power in pulse trains, not sinusoidal waveforms. Figure 6.21 illustrates the operation of a simple VFD.

On the left side, the line current is directed through a series of diodes. These diodes perform *full-wave rectification* on the input power. This means the input sinusoid, which ranges from positive to negative, is converted into a series of half-sinusoids whose values are always positive. This rectified power is connected across a capacitor, which reduces the ripple and produces near-DC power.

On the right side, the rectified DC power is passed into a series of MOSFETs that serve as electrical switches. As discussed in Chapter 2, when voltage is applied to a MOSFET's gate, the resistance between its source and drain becomes negligible. In this circuit, called an *inverter*, the upper MOSFETs connect the upper voltage to the motor's output and the lower MOSFETs connect the lower voltage.

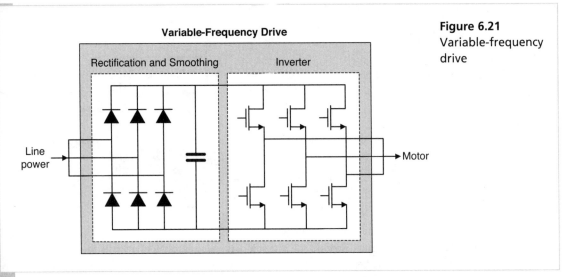

Figure 6.21
Variable-frequency drive

A microprocessor is connected to each of the MOSFETs' gates. Applying voltage to different gates generates pulses to be sent to the motor. These pulses are formatted according to pulse width modulation (PWM), which increases power to the motor by increasing the pulse width.

VFDs commonly deliver power using sinusoidal PWM, or SPWM. This method increases and decreases the pulse width in a sinusoidal fashion. Figure 6.22 shows how this works.

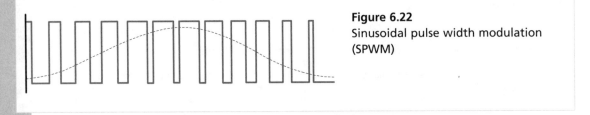

Figure 6.22
Sinusoidal pulse width modulation (SPWM)

The sinusoid's frequency is determined by the VFD settings. In many cases, it's important that the motor provide maximum torque regardless of the frequency. For this reason, many VFDs keep the voltage/frequency ratio constant.

6.6.3 VFD Harmonic Distortion

VFDs provide many advantages, but one disadvantage involves the waveform of the output power. No matter how good the VFD's rectifier is, its output power will still contain traces of the original input AC power. To be specific, its output will contain frequencies at multiples of the original frequency. These frequencies are called *harmonics*, and having harmonics present in the VFD's output can make it difficult to control a motor with precision. For example, if the original AC frequency is 60 Hz, its harmonic frequencies are 120 Hz, 180 Hz, 240 Hz, and so on.

One significant issue caused by harmonics is noise. Harmonic distortion can cause minor deviations in the motor's operation that produce sound. In addition, the high-frequency components in the power can interfere with electric circuits and RF communication. For this reason, it's important to keep VFDs and high-speed motors away from critical circuitry.

With modern VFDs, harmonic distortion isn't a significant problem. However, if power from a regular VFD won't be sufficient, you can get better results from a *multipulse* VFD. This type of VFD uses multiple rectification stages.

A 6-pulse VFD rectifies the AC power with the six-diode rectifier presented in Figure 6.21. A 12-pulse VFD uses two rectification stages, and an 18-pulse VFD uses three. These additional stages significantly reduce the amount of harmonic frequency content in the VFD's output. The drawback of multipulse VFDs is that they're larger and more expensive than regular VFDs.

Other methods for reducing harmonic content include low-pass broadband filtering and placing an inductor in series with the VFD's power. An inductor's impedance increases with frequency, and an inductor used to reduce high-frequency content from a signal is called a *DC choke*.

6.7 Universal Motors

No discussion of AC motors would be complete without mentioning *universal motors*, which can run off of DC power or single-phase AC power. These are commonly found in appliances such as food processors and vacuum cleaners.

In essence, a universal motor is a series-wound brushed DC motor whose structure is slightly modified to receive AC power. As discussed in Chapter 3, a series-wound brushed DC motor has the following properties:

- The rotor's windings are connected in series to the windings in the stator, which generate a magnetic field.

- The rotor's windings receive power through a mechanical commutator called a brush.

The problem with running DC motors on AC power is that the windings in the rotor and stator add a great deal of inductance to the circuit. This increases the phase difference between voltage and current to nearly 90°. This reduces the power factor to cos(90°), which equals 0. In other words, the voltage and current are so out of phase that the motor can't perform real work.

To help solve this problem, the universal motor has a *compensating winding* connected in series. This reduces the inductance in the armature and thereby reduces the phase difference between the current and voltage. Figure 6.23 presents a simplified circuit for the universal motor.

In addition, the number of wires in the stator's windings is kept to a minimum. This reduces the stator's inductance in a universal motor.

One important advantage of universal motors is their high starting torque. Because the rotor and stator windings are connected in series, an increase in current increases both magnetic fields. Therefore, the motor's high starting current produces a significant starting torque.

The main drawback of universal motors, like that of brushed DC motors, is the brush. The brush's presence reduces the motor's efficiency, and its operation produces friction and heat that degrade the motor over time.

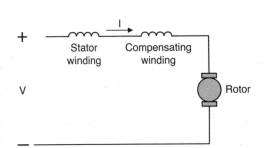

Figure 6.23
Universal motor circuit

6.8 Summary

DC motors are more common in maker projects, but AC motors are more numerous throughout the world. Nearly all the electric motors used in industry and households run on AC current, so it's a good idea to understand how they work. The goal of this chapter has been to present different AC motors so you can select the most suitable motor for your projects.

There are many different types of AC motors, so it's important to classify them according to their operation. Polyphase motors run on AC power with multiple phases (usually three), whereas single-phase motors run on power with a single phase. In both cases, the input power is delivered to windings in the stator, which produce a rotating magnetic field.

The speed of a motor's rotating field is called its synchronous speed. A synchronous motor rotates at the same speed as its synchronous speed. Asynchronous motors rotate slightly slower.

In an asynchronous motor, the rotor has conductors that receive induced voltage from the rotating field. This induction causes the rotor to turn at a speed less than that of the field's rotation, and the difference between the motor's speed and the synchronous speed is called slip. The torque-speed properties of a squirrel-cage induction motor can't be changed because the rotor's conductors can't be accessed. In contrast, the conductors of a wound-rotor have variable resistance. Increasing resistance reduces the starting current and increases the starting torque.

This chapter presented three types of synchronous motors: doubly excited synchronous motors, permanent magnet synchronous motors, and synchronous reluctance motors. The only difference between them is the rotor. The rotor of a doubly excited synchronous motor receives DC current, and the rotor of a permanent magnet synchronous motor has magnets mounted on its perimeter. The rotor of a synchronous reluctance motor is made of a ferromagnetic material that follows the rotating field.

In a single-phase motor, the stator needs special circuitry to generate a rotating field. Split-phase motors have two windings in the stator, a main winding and an auxiliary winding, but the phase difference is too small to generate a significant field. For this reason, capacitor-start and capacitor-run motors have a capacitor in series with the auxiliary winding. In a shaded-pole motor, the stator's windings have a blocked-off element encircled by a conductive shading coil. This is sufficient to generate a rotating field, but these motors are only sufficient for low-torque applications.

AC motors usually receive power directly from household/industry outlets, so motor control isn't a significant concern. However, this chapter presented two methods of AC motor control: eddy-current drives and variable-frequency drives. Eddy-current drives transform the shaft output of one motor to the output of another shaft. Variable-frequency drives receive power of one voltage and frequency and provide power with another voltage and frequency.

Universal motors can run on DC or single-phase AC power. They're essentially series-wound brushed DC motors whose inductance is reduced for better AC performance. They're inexpensive and provide high starting torque, but the presence of the brush makes them inefficient and prone to mechanical failure.

7

GEARS AND GEARMOTORS

If you take an ordinary DC motor from a hobby shop and apply power to its leads, you'll see that the shaft rotates quickly—hundreds or thousands of rotations per minute. But if you attach a significant load to the shaft, it may not turn at all. This is a frequent problem for makers building motorized systems: The motor rotates at high speed but can't exert the necessary amount of torque.

A common solution is to buy special motors called *gearmotors*. A gearmotor is an integrated combination of an electric motor and a gear. The gear increases the torque delivered to the load and reduces the motor's speed. Now the problem is to select the right gearmotor. Should you buy the 6:1 spur gearmotor or the 26:1 planetary gearmotor? What about the gearmotor that ensures low noise by combining a worm gear and a planetary gear?

The goal of this chapter is to give you the information you need to understand gearmotors and choose the right one for your project. To introduce this subject, the chapter begins with an overview that discusses mechanical advantage and the many types of gears available. For each gear, I'll explain how it works and present its advantages and disadvantages. Once you understand the different types of gears, the topic of gearmotors becomes straightforward.

7.1 Overview of Gears

A *gear* is a toothed element that can be connected to a motor's shaft. Its teeth interact, or *mesh*, with other toothed elements to change how the motor's torque, speed, or direction is applied to the load. The first part of this chapter explains how this change in motion works. Then we'll look at two aspects of gears that every designer should be familiar with: pitch and backlash.

7.1.1 Power Transmission

As discussed in Chapter 2, "Preliminary Concepts," a motor's power equals its torque multiplied by angular speed. Put mathematically, $P = \tau\omega$.

Suppose that a motor's shaft is attached to a gear (the *input gear*) that meshes with a second gear (the *output gear*). Ideally, the input gear transmits all of its power to the output gear. That is, $\tau_o\omega_o = \tau_i\omega_i$.

This relationship doesn't imply that the two gears rotate at the same speed or that they exert the same torque. Consider the gears depicted in Figure 7.1. The input gear has six teeth along its perimeter and the output gear has 10.

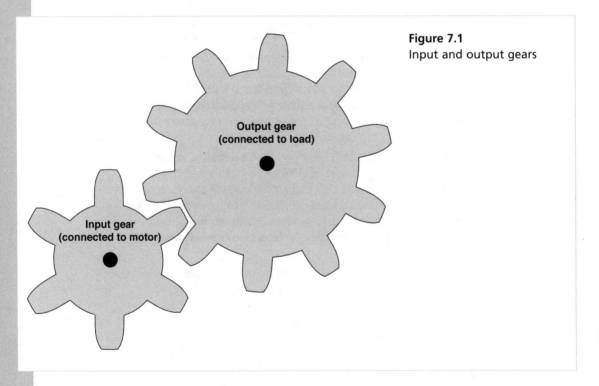

Figure 7.1
Input and output gears

Output gear
(connected to load)

Input gear
(connected to motor)

By the time the input gear completes a revolution, the output gear completes only six-tenths of its revolution. Therefore, the input gear's rotational speed is greater than that of the output gear. Put mathematically, $\omega_o = 0.6\omega_i$. Denoting the number of teeth on the input gear as N_i and the number of teeth on the output gear as N_o, the relationship between angular speed and the number of teeth can be expressed with the following expression:

$$\frac{\omega_o}{\omega_i} = \frac{N_o}{N_i}$$

The output gear is slower than the input gear, but the mechanical power ($\tau\omega$) is the same for both. Therefore, the output gear must exert greater torque than the input gear. The torque ratio equals the speed ratio, so we can extend the preceding equation as follows:

$$\frac{\tau_o}{\tau_i} = \frac{\omega_i}{\omega_o} = \frac{N_o}{N_i}$$

In practical motor systems, it's common to connect a motor's gear to a larger gear to increase torque. This is called *gear reduction* because the output speed is less than the input speed. The reduction is usually expressed as *X*:1, where *X* is the proportional increase in torque. This increase in torque is frequently referred to as the gear's *mechanical advantage*. It's also common to see the term *gear ratio* used in place of *gear reduction*.

For example, a 3:1 gear produces three times the input torque and one-third the speed. A 4:1 gear produces four times the input torque and one-fourth the speed. Gears can be combined to produce greater reduction. If a 5:1 gear is connected to a 6:1 gear, the resulting torque will be 6×5 = 30 times greater than the input torque. A combination of gears is called a *gear train*.

As gears interact, the contact friction reduces the power transmitted to the output gear. This power loss is expressed in terms of efficiency. A gear's efficiency is defined by the following equation:

$$\eta_{gear} = \frac{Power_{output}}{Power_{input}} = \frac{\tau_o \omega_o}{\tau_i \omega_i}$$

Gears are generally very efficient, with η_{gear} values typically between 90% and 98%. As a later section will show, some types of gears are more efficient than others.

Chapter 2 explained that the efficiency of an electric motor equals the mechanical power ($\tau_i \omega_i$) divided by the electrical power (VI). If a system contains an electric motor connected to gears, the system's efficiency can be determined with the following equation:

$$\eta_{system} = \eta_{motor} \cdot \eta_{gear} = \left(\frac{\tau_i \omega_i}{VI}\right)\left(\frac{\tau_o \omega_o}{\tau_i \omega_i}\right) = \frac{\tau_o \omega_o}{VI}$$

In this case, the system's efficiency equals the product of the efficiencies of the system's components. Multiple gear stages can provide significant gear reduction, but they can also substantially reduce the system's efficiency.

7.1.2 Pitch

Gears are simple to understand, but the terminology can be confusing. For example, one spur gear's characteristics are given simply as "48P 90T." The "T" stands for the number of teeth, so the gear has 90 teeth along its perimeter.

The "P" stands for pitch, and when we're discussing gears, there are two types of pitch: circular pitch and diametral pitch. To help make sense of this, Figure 7.2 presents the same two gears as Figure 7.1, but adds markings that describe pitch.

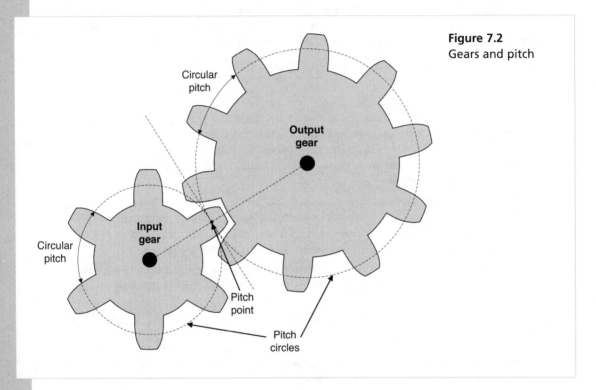

Figure 7.2
Gears and pitch

The center of the figure contains two straight lines: one between the centers of the two gears and one that corresponds to the gears' direction of motion. The intersection of these two lines is called the *pitch point*.

Both gears are drawn with a dashed circle that passes through the pitch point. This circle is called the *pitch circle*. The diameter of a gear's pitch circle is its pitch diameter. The pitch radius is half of the pitch diameter.

A gear's *circular pitch* is the distance between a point on one tooth and a corresponding point on an adjacent tooth. Circular pitch is measured along the pitch circle, and if two gears have different circular pitches, they can't mesh properly.

A gear's *diametral pitch* is the ratio of the number of teeth to the pitch diameter. If you purchase a gear, this is the value identified with P. For example, if a gear is sold as 48P, it has 48 teeth per inch of pitch diameter. Other common values are 24P and 32P. As with circular pitch, if two gears have different diametral pitches, they can't mesh properly.

7.1.3 Backlash

In an ideal system, the teeth of two meshed gears would fit so closely together that every motion of the input gear moves the output gear. Also, if the input gear changes direction, the output gear changes direction immediately.

In practical systems, there's always an extra amount of space between the gears' teeth. This means that small movements of the input gear may not affect the motion of the output gear. Also, if the input gear changes direction, it will take time for its teeth to contact those of the output gear. This lost motion is referred to by a number of names, including *slop* and *play*. However, the most common term I've encountered is *backlash*.

A certain amount of backlash is tolerated to reduce the possibility of jamming and allow for lubrication, thermal expansion, and minor variations in tooth thickness. To provide backlash, gear designers make the size of the space between teeth larger than the width of a tooth. Put another way, they make the circular pitch greater than twice the width of a tooth. Another method of increasing backlash involves moving the gears further apart.

Backlash can be computed mathematically by subtracting the width of a tooth from the width of the gap between adjacent teeth. Both dimensions are measured along the gear's pitch circle.

7.2 Types of Gears

A gear's purpose is to change the torque, speed, and/or direction delivered by a power source to a load. The first gears were invented in medieval times, and since then, hundreds of different types of gears have been constructed. This section won't present them all, but focuses on six of the most common gears: spur gears, helical gears, bevel gears, racks and pinions, worm gears, and planetary gears.

7.2.1 Spur Gears

The oldest and simplest type of gear is the spur gear, which is a disk with teeth extending from its perimeter. These are the gears depicted in Figures 7.1 and 7.2. The shafts attached to spur gears are always parallel to one another.

The principal characteristics of a spur gear are its pitch and number of teeth. This is given by XP YT, where X identifies the diametral pitch (number of teeth per inch of diameter) and Y is the number of teeth. As an example, Figure 7.3 presents a 48P 76T spur gear from Traxxas.

One drawback of using spur gears is vibration. As shown in Figure 7.2, only one or two pairs of teeth are in mesh at any time. This number is called the gear's *contact ratio*, and for spur gears, the value is generally between 1.2 and 2. Also, the teeth are either fully in contact or not touching at all.

As the gears' speeds increase, the abrupt changes in contact produce vibration. These vibrations wear down the gears and affect the stability of the overall system. The most noticeable effect of gear vibration is noise, which becomes particularly important in automotive systems.

In addition to vibration, the changes in contact produce significant stress in the gears' teeth. This increases the possibility of breakage and limits the amount of power that can be transmitted from one shaft to another.

Figure 7.3
A 48P 76T spur gear

7.2.2 Helical Gears

To reduce the vibration associated with spur gears, engineers designed gears whose teeth come into contact in a more gradual manner. The teeth of these gears are oriented at an angle called the *helix angle*. Figure 7.4 shows what the HL20L helical gear from Boston Gear looks like.

Figure 7.4
The HL20L helical gear

Because the teeth are slanted, more pairs of teeth can be in mesh at once. The contact ratio of a helical gear is usually between 2.2 and 4. As a result, the power transmission between two helical gears is smoother than that between two spur gears, producing less vibration and less noise.

A disadvantage of using helical gears is that some of the force imparted on a tooth is exerted perpendicularly to the direction of rotation. This axial force depends on the gear's helix angle, and Figure 7.5 makes this relationship apparent.

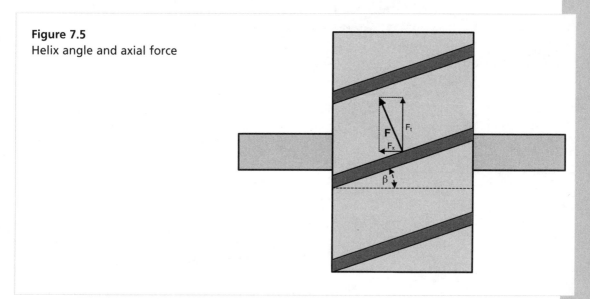

Figure 7.5
Helix angle and axial force

In this figure, the helix angle is denoted as β. If F is the total force exerted on a tooth with angle β, the axial force equals $F \sin(\beta)$ and the tangential force equals $F \cos(\beta)$.

To manage the axial force, a system with helical gears needs thrust bearings to prevent damage to the system. Also, because of the friction involved, helical gears can generate a significant amount of heat. This makes helical gear systems less efficient than systems with spur gears.

To remove the axial force associated with slanted teeth, helical gears have been developed with teeth slanted in both directions. These are commonly called *herringbone gears*, and Figure 7.6 gives an idea of what they look like.

In most cases, helical gears are connected to parallel shafts. But helical gears can also transmit power between shafts that aren't parallel. In this case, the helical gears are called *crossed gears*. If the helical gears' shafts are perpendicular, the gears are called *skew gears*.

7.2.3 Bevel Gears

Just as a spur gear is a cylinder with teeth on its outer surface, a bevel gear is a truncated cone with teeth on its outer surface. Figure 7.7 gives a basic idea of what this looks like.

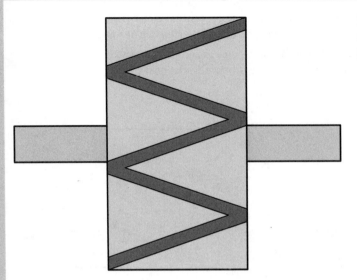

Figure 7.6
Dual-directional teeth of a herringbone gear

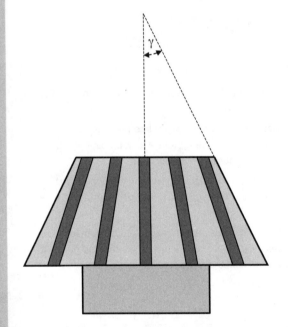

Figure 7.7
Geometry of a bevel gear

In this figure, the angle γ is referred to as the gear's *pitch angle*. If two bevel gears are in mesh, the angle between their shafts equals the sum of the gears' pitch angles. In most cases, the pitch angle of a bevel gear is 45° and the shafts of meshed bevel gears are perpendicular. Figure 7.8 shows what a real-world coupling of bevel gears looks like.

Figure 7.8
Bevel gears oriented at a right angle

If a bevel gear's teeth are straight, as in a spur gear, it's referred to as a straight bevel gear. If its teeth are curved, it's called a spiral bevel gear. As with the teeth of a helical gear, these curved teeth increase the contact ratio between two meshed gears, ensuring smooth rotation with minimal vibration. Both gears depicted in Figure 7.8 are spiral bevel gears.

One type of spiral bevel gear has a curved overall shape instead of a conical shape. More precisely, the gear's shape is that of a hyperbola, and for this reason, the gear is called a *hypoid gear*.

7.2.4 Rack and Pinion

Just as a bevel gear changes how rotation is oriented, a rack and pinion converts rotational motion to linear motion. The pinion is a spur gear or helical gear, and the rack gear is a straight bar with teeth that mesh with the pinion's. You can think of a rack gear as a spur/helical gear with infinite radius.

In many systems, the pinion is the input gear and the rack is the output gear. Figure 7.9 shows what a rack and pinion looks like.

One important application of rack and pinion gears involves automobile steering. When the steering wheel turns, its shaft rotates a pinion connected to a rack gear. As the rack moves left and right, it moves a tie rod that controls the angles of the vehicle's wheels.

In addition to changing rotational motion to linear motion, the rack and pinion in a steering system also provides gear reduction. This is why, before the advent of power steering, it took multiple rotations of the steering wheel to change the tires' angles from fully left to fully right.

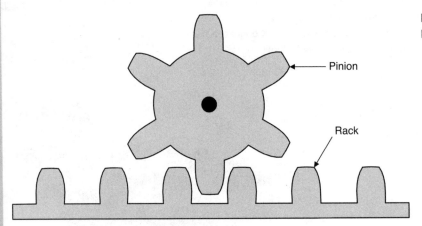

Figure 7.9
Rack and pinion gear

Pinion

Rack

7.2.5 Worm Gear

Worm gears can be hard to understand at first, so I'll compare them to screws. Like a screw, a worm gear is long and narrow, with a thread wrapped around its outer surface. Unlike a screw, a worm gear doesn't taper to a point. Instead, its overall shape is cylindrical with a constant radius.

When a carpenter uses a screw, he/she rotates the screw, causing the thread to drive further into the receptacle. A worm gear works in a similar manner. As the worm gear rotates, its thread turns the teeth of a second gear, which is usually a spur gear or helical gear.

Like a bevel gear, a worm gear changes the orientation of the input rotation by 90°. Figure 7.10 shows a worm gear meshing with a spur gear.

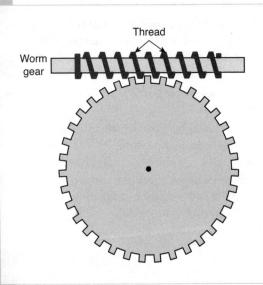

Thread

Worm
gear

Figure 7.10
A worm gear

Each time the worm gear completes a revolution, it advances a single tooth of the meshed gear. Therefore, the gear reduction depends on the number of teeth in the meshed gear. With worm gears, it's common to see gear reductions of 20:1 up to 100:1 and 300:1. This significant gear reduction in a confined space is the most important advantage of worm gears.

A worm drive only transmits power in one direction. That is, in Figure 7.10, the turning worm gear rotates the spur gear, but rotating the spur gear will not rotate the worm gear. Put another way, the input shaft connected to the worm gear is not affected by the output shaft.

As a worm gear rotates, its teeth remain in constant contact with the teeth of the meshed gear. This ensures silent operation with low vibration, but also produces heat through friction. This loss of power means that worm gears tend to have low efficiency compared to other gear drives, and the efficiency decreases as the gear reduction increases.

7.2.6 Planetary Gear

A planetary gear provides significant gear reduction, and unlike any gear mentioned so far, the output shaft has the same direction and center as the input shaft. This means there's no need for side-shafts to bear weight.

A planetary gear consists of a meshed arrangement of simpler gears, and for this reason, it's commonly referred to as a gear train instead of a gear. Figure 7.11 illustrates the internal structure of a planetary gear.

Figure 7.11
Structure of a
planetary gear

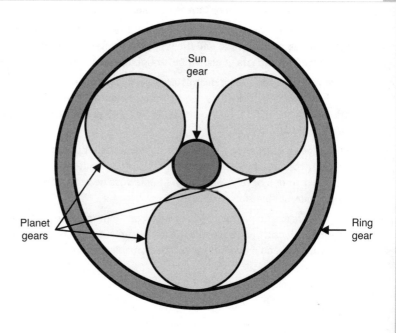

A planetary gear contains two gears with the same center: the sun gear on the inside and the *ring gear* on the outside. Between the two gears, it has two or more additional gears called *planet gears.* Each planet gear is in mesh with the sun gear and the ring gear.

The planet gears are generally connected to one another through a single rotating element called the *carrier* or *cage.* As a result, they rotate uniformly around the sun gear. In this manner, the planetary gear has three potentially moving parts: the ring gear, the sun gear, and the planet carrier. All three parts rotate around the same center.

 note

I've omitted the teeth of the gears in Figure 7.11, but the gears that make up a planetary gear are usually spur gears or helical gears.

The gear reduction depends on which gear is used for input and which is used for output. One common configuration connects the input shaft to the sun gear and the output shaft to the planet carrier. This not only provides significant gear reduction, but also allows the sun gear to distribute torque among the planet gears. This enables planetary gears to be used in high-torque applications that might break a regular gear train.

7.3 Gearmotors

If you're building a motorized system, you don't need to work directly with spur gears or worm gears. It's more common to purchase a motor with one or more integrated gears. These gearmotors aren't necessary if the application involves rotating the wheels of an RC vehicle, but if the motor needs to lift a robotic arm or turn a propeller through water, they become essential.

A wide variety of gearmotors is available on the market. Besides cost, the main differences between them involve the gear ratio and the type of the gear integrated into the device. If you have a solid understanding of gears, it should be straightforward to find a suitable option for your application.

From what I've seen, most gearmotors rely on spur gears or helical gears for power transmission. They can be distinguished by the center of the output shaft, which is parallel to the motor's cylindrical body but offset from the center. Figure 7.12 presents an example: the 29:1 gearmotor from SainSmart.

In contrast, if a gearmotor's shaft has the same center as the system's cylindrical body, the motor probably contains a planetary gear. If the motor's shaft is perpendicular to the body, it may contain a worm gear or a bevel gear. Maxon Motors sells gearmotors that integrate a planetary gear and a worm gear into the same package.

Gearmotors are frequently used in robotic and high-precision applications, so many gearmotors have integrated sensors and encoders. As an example, the gearmotor depicted in Figure 7.12 has Hall effect sensors and a quadrature encoder capable of resolving angles down to $360°/64 = 5.625°$.

Figure 7.12
The SainSmart 29:1 gearmotor

7.4 Summary

When I started working with electric motors, I was surprised by how little torque they provide. Because they're so heavy and consume so much current, I assumed that they provide a great deal of torque. This isn't the case, but the motor's torque can be increased using gears.

A gear's torque increase is expressed as X:1, where X is the ratio of the output torque to the input torque. This factor, referred to as gear ratio or gear reduction, depends on the ratio of the number of teeth on the meshed gears. If the input gear has N_i teeth and the output gear has N_o teeth, the output torque will be N_o/N_i times as large as the input torque.

In addition to the number of teeth, gears are commonly classified by pitch. The pitch circle is the imaginary circle surrounding a gear that intersects the point of action. Circular pitch is the distance between similar points on adjacent teeth, as measured along the pitch circle. Diametral pitch equals the gear's number of teeth divided by the diameter of its pitch circle. Unless two gears have equal circular and diametral pitches, they can't mesh.

The majority of this chapter has focused on presenting the different types of gears. The simplest and most common gear is the spur gear, which is a disk with teeth protruding from its outer surface. The interaction of spur gears can produce vibration and noise, so helical gears have teeth slanted at an angle. This ensures smoother, quieter operation, but a portion of the transmitted force is directed axially, which reduces efficiency.

Bevel gears and worm gears change the direction of the applied power. That is, these gears make it possible for the output shaft to have a different angle than the input shaft. In general, worm gears provide significantly more gear reduction than bevel gears, but because a worm gear's teeth are always in contact, its operation can produce heat that reduces the system's efficiency.

A combination of gears is called a gear train, so technically, both the rack and pinion and planetary gears are gear trains. The rack and pinion converts rotational motion to linear motion (or vice versa). The planetary gear contains multiple rotating elements, all rotating about the center. The gear reduction depends on which element is connected to the input and which is connected to the output.

A gearmotor is an electric motor with one or more integrated gears. Once you have a solid grasp of the different types of gears available, it's straightforward to select a gearmotor for your project. But keep in mind that gears always reduce efficiency, and because of the mechanical contact, gears always add heat to the system and the possibility of damage.

LINEAR MOTORS

Until this chapter, every electric motor discussed in this book has produced rotary motion. Voltage and current go in, torque and rotational speed come out. This chapter leaves those motors behind and takes a look at linear motors. These motors move in a straight line with linear force and speed.

Linear motion is needed in many types of systems, but linear motors aren't commonly used by makers or professional engineers. That's because these motors aren't well-understood, there aren't many manufacturers, and they're expensive. For these reasons, designers frequently obtain linear motion by connecting rotary motors to mechanical elements. As an example, the popular 3D printer RepRap obtains linear motion by connecting stepper motors to a timing belt.

Despite their rarity, linear motors are worth studying. When it comes to linear motion, linear motors provide better speed and precision than comparable electromechanical linkages—and depending on the motor, they can make more efficient use of power.

One major application of linear motors is transportation. Electric trains rely on linear motors to carry passengers and freight across long distances. Nations have spent billions on maglev (magnetic levitation) trains, and much of this effort has been devoted to linear motor research.

This chapter presents an overview of the topic of linear motors, dividing them into four categories:

- Linear actuators

- Linear synchronous motors

- Linear induction motors

- Homopolar motors

In addition to discussing the structure and operation of linear motors, this chapter presents a number of particularly interesting applications. These include coilguns, rail guns, and maglev trains. These futuristic technologies are actively researched across the globe, and someday, they may cross the border from science fiction to commonplace usage.

8.1 Linear Actuators

Chapter 2, "Preliminary Concepts," explained the fundamentals of electromagnets. If a current-carrying wire is wrapped around an iron core, the wrapped core will behave like a magnet, with north and south poles. If current is reversed, the north and south poles switch sides. If current is turned off, the magnetic behavior stops.

If an iron core is only partially inserted into a current-carrying coil of wire, an interesting thing happens. The iron core (called a *plunger*) experiences a force that draws it fully into the coil. Figure 8.1 shows what this looks like.

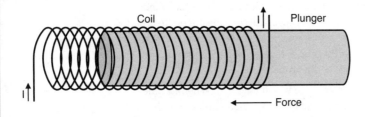

Figure 8.1
Motion of an electrical linear actuator

This type of device is referred to by many names, including solenoid, solenoid actuator, electric actuator, magnetic actuator, and linear actuator. Some sources refer to it as a linear motor, but from what I've seen, it's more commonly called an actuator than a motor. For this reason, I'll refer to it as a *linear actuator*.

This raises an important question—what's the difference between an actuator and a motor? The answer is subtle and relates to the device's function, not its structure. A motor is a device that converts energy into motion. An actuator is a specific type of motor whose motion is intended to control another mechanism.

For example, one common application of a linear actuator is to control the position of a mechanical switch. In this case, the actuator is called a *relay*. Other actuators open and close valves. In contrast, if a device's purpose is simply to effect motion, such as spinning the wheels of a remote-controlled car or rotating the propellers of a quadcopter, it's a motor, not an actuator.

Another difference between linear actuators and linear motors involves the nature of the input power. Every linear actuator I've encountered requires DC power, but as later sections will show, linear synchronous motors and linear induction motors rely on AC power.

8.1.1 Operation and Structure

To see how a linear actuator works, it's important to grasp the concept of *magnetic energy*. Preceding chapters have explained how magnets and electromagnets perform mechanical work. A device's magnetic energy is the measure of how much work its magnets are capable of performing.

If current flows through an empty coil of wire, the region inside the coil contains a small amount of magnetic energy. However, if an iron plunger is fully inserted into the coil, the coil's energy increases substantially. If the plunger is partially inserted, the region containing the plunger has more energy than the region that doesn't.

The force on the plunger equals this difference in energy divided by the distance between the plunger and the end of the coil. The mathematical relationship for the force is complicated, but there are three important relationships to be aware of:

- Force increases with the square of the input current.
- Force increases with the square of the number of turns in the coil.
- Force increases with the cross-sectional area of the coil.

The force always pulls the plunger into the coil, but linear actuators can be constructed to push or pull the intended load. Figure 8.2 presents a linear actuator that pulls the load using a hook.

Figure 8.2
Linear actuator (pull-type)

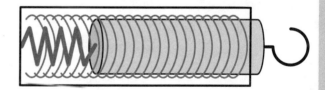

The metal casing concentrates the coil's magnetic field inside the actuator. The plunger is attached to the casing with a spring. When current is sent through the coil, the force pulls in the plunger and compresses the spring. When the current is removed, the spring extends, returning the plunger to its original position.

Figure 8.3 depicts a linear motor that pushes the intended load. Here, the right side of the plunger is attached to a non-ferromagnetic head that pushes outward when the coil is activated.

When current is delivered to the coil, the force pushes the plunger into the coil. When current is removed, the spring contracts and returns the plunger to its original position.

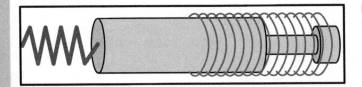

Figure 8.3
Linear actuator (push-type)

8.1.2 Sample Linear Actuator

Figure 8.4 depicts a linear actuator sold by Amico. This is a pull-type linear actuator, and when power is supplied, the plunger retracts 5 mm.

Figure 8.4
Linear actuator from Amico

The electrical properties of this actuator are given as follows:

- **Required voltage**: 5 V DC
- **Maximum current draw**: 1.1 A

Actuators such as this are commonly used to open doors and hatches in vehicles. Smaller actuators are used in power door locks in cars. Push/pull actuators lock and unlock the car door when driven with an electric signal.

8.1.3 Coilguns

In most linear motors, the plunger is attached to a mechanism so that it can be pushed or pulled repeatedly. However, if the plunger isn't attached and the coil's current is sufficiently strong, the motor can launch the plunger as a projectile. In this case, the plunger is referred to as a *sabot* and the motor is called a *coilgun*.

Compared to regular linear actuators, coilguns require a great deal of power. For a sabot the size of a thimble, a coilgun might require as much as 30 V at 15 A, but this power is only required for the tens or hundreds of milliseconds needed to propel the sabot out of the coil. Because of the large power requirements over a short time, many designs rely on a bank of large capacitors to store energy. By discharging the capacitors simultaneously, the power circuit can supply enough electricity to launch the sabot.

Coilguns have fascinated electrically minded amateurs and professionals since the 1930s. A casual Internet search on the topic will lead to many hobbyist sites that discuss physics, parts, and circuit designs. However, due to the large power draw, coilguns have never become practical tools.

This doesn't mean coilguns won't be important in the future. As I write this, the United States military is investigating the use of coilguns to launch aircraft and mortar rounds from naval carriers. NASA is looking into coilgun-based launchers for its satellites and rockets.

8.2 Linear Synchronous Motors

For rotary AC motors, asynchronous (induction) motors are more common than synchronous motors. But when it comes to linear motion, linear synchronous motors, abbreviated LSMs, are more common.

Chapter 6, "AC Motors," presented three different types of synchronous AC motors: doubly excited motors, permanent magnet motors, and reluctance motors. The vast majority of LSMs I've seen have permanent magnets, so this section focuses exclusively on permanent magnet LSMs. To be specific, this chapter discusses the structure of permanent magnet LSMs and then presents two real-world examples: the Yaskawa SGLG motors and the Transrapid maglev train line.

8.2.1 Structure

In general, linear motors are constructed by unrolling rotary motors. Figure 8.5 shows what I'm talking about. The top portion presents a rotary synchronous motor and the bottom portion shows what the unrolled LSM looks like.

Like rotary synchronous motors, LSMs receive AC power, usually split into three phases. But the terms used to describe the elements of an LSM are different:

- The moving element is called the *forcer*, not the rotor.

- The stationary element is called the *rail* or *track*, not the stator.

In the figure, the magnets are on the forcer and the windings are on the track. This is called a *long stator* design. In a *short stator* design, the windings are on the forcer and the magnets are on the track. The following discussion presents three short stator motors: iron-core LSMs, ironless LSMs, and slotless LSMs. But first, I'd like to explain how synchronous speed is computed for linear motors.

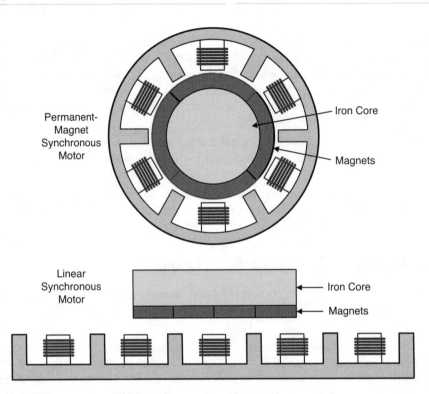

Figure 8.5
Rotary and linear synchronous motor

Permanent-
Magnet
Synchronous
Motor

Iron Core

Magnets

Linear
Synchronous
Motor

Iron Core

Magnets

Linear Synchronous Speed

Chapter 6 explained how one cycle of AC power corresponds to a single rotation of the magnetic field inside the stator. If the input power has frequency f, the time of the rotation is 1/f.

For linear motors, the situation is similar. One cycle of AC power corresponds to a change in the magnetic field that travels from Winding A back to Winding A. As shown in Figure 8.6, this distance equals 2τ.

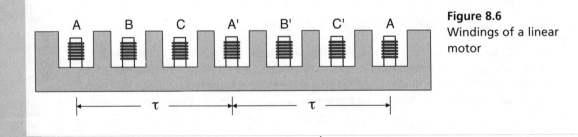

Figure 8.6
Windings of a linear motor

If the input power has frequency f, the time required to travel across a cycle is 1/f. Therefore, the motor's *linear synchronous speed* is given by the following equation:

$$v_s = \frac{distance}{time} = \frac{2\tau}{1/f} = 2\tau f$$

The operating speed of an LSM equals this value. For linear motors, it's given in units of meters/second instead of revolutions per minute. Also, the equation doesn't depend on the number of poles. This is because an increase in the number of poles produces a corresponding increase in τ.

Iron-Core LSM

An iron-core LSM is essentially similar to the linear motor shown in Figure 8.5, but the windings are on the forcer and the magnets are on the rail. Figure 8.7 shows what this looks like.

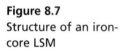
Figure 8.7
Structure of an iron-core LSM

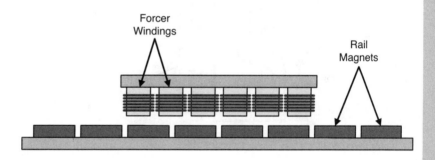

This is referred to as an iron-core LSM because the forcer's windings are wrapped around iron cores. This increases the amount of force the motor can deliver, which increases the maximum load that it can push, pull, or carry.

One drawback of iron-core LSMs involves the forcer's weight and the iron's attraction to the rail magnets. Because of the weight and magnetic attraction, the motor designer must take careful steps to ensure that the forcer doesn't come in contact with the rail. This usually involves a sliding mechanism that keeps the forcer at least 0.75 mm away.

Because of the iron's attraction to the rail magnets, the forcer may be reluctant to pass from one magnet to the next. This phenomenon, called *cogging*, produces a jerky or rippling motion as the forcer moves. It can be reduced by moving the forcer at high speed and orienting the magnets at an angle.

Ironless LSM

As its name implies, an ironless LSM doesn't have iron cores in its forcer. The windings are wrapped around air, and for this reason, ironless LSMs are commonly referred to as *air core* motors.

Without the weight and magnetic attraction associated with iron, the forcer can move smoothly across the track, without cogging or attraction to the rail's magnets.

The primary drawback of ironless LSMs is that the forcer can only exert a small amount of force. For this reason, iron-core LSMs are used in applications involving high loads and ironless LSMs should be used when precision and speed control are more important.

To increase the force that can be exerted, the track of an ironless LSM has two rows of permanent magnets. These magnets surround the forcer, which is mounted vertically. Figure 8.8 depicts a cross-section of a typical ironless LSM.

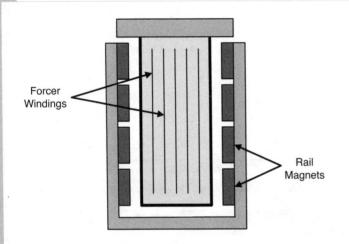

Figure 8.8
Cross-section of an ironless LSM

Forcer Windings

Rail Magnets

Another disadvantage of using ironless LSMs is cost. Permanent magnets are expensive, especially the rare-earth magnets commonly used in linear motors. Ironless LSMs have two rows of magnets, and this makes them significantly more expensive than iron core LSMs. The expense increases as the length of the track increases.

The enclosed structure of the ironless LSM makes it difficult to dissipate heat. For this reason, ironless motors can't be run at the same current levels as an iron-core motor. In addition, it's common to have heat sensors on the forcer to ensure that the motor isn't damaged during operation.

Slotless LSM

A slotless LSM serves as a compromise between an iron-core and an ironless LSM. Like an iron-core motor, its track has a single row of magnets and the forcer is positioned on top of the track. Like an ironless motor, the windings are wrapped around air. Figure 8.9 gives an idea of what this looks like.

Slotless motors can't support the same high loads as iron-core motors, but because of the iron backing, they can exert more force than ironless motors. They also have better heat dissipation than ironless motors because the forcer travels on top of the track. Further, the track has only one row of magnets, which makes it less expensive than motors with two rows of magnets.

Figure 8.9
Cross-section of a slotless LSM

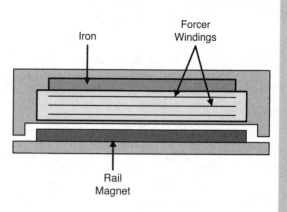

The main drawback of slotless LSMs involves efficiency. Slotless motors don't have iron-core windings or multiple rows of magnets. Therefore, to obtain the same degree of motion, a slotless motor requires more power than a comparable iron-core or ironless motor.

8.2.2 Case Study: Yaskawa SGLG

The Yaskawa Electric Corporation manufactures a number of different types of AC motors, including induction motors and servomotors. Yaskawa also provides drive systems for elevators, pumps, and air conditioning systems.

Yaskawa manufactures three different lines of linear synchronous motors, each with a different four-letter designation. The SGLF and SGLT lines of motors are iron-core motors, with forcers that travel on top of a track. The SGLG line consists of ironless motors with two rows of rare-earth magnets on either side of the forcer. Figure 8.10 depicts one of the motors in Yaskawa's SGLG line of motors.

Figure 8.10
The Yaskawa SGLG linear synchronous motor

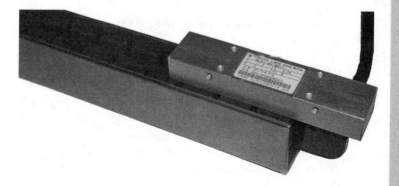

Table 8.1 compares the characteristics of two motors in the SGLG and SGLF lines. The SGLFW-35A is an iron-core LSM and the SGLGW-40A is an ironless LSM.

Table 8.1 Comparison of Yaskawa's Iron-Core and Ironless LSMs

LSM Characteristic	SGLFW-35A	SGLGW-40A
Forcer windings	Iron-core	Ironless
Continuous force	80N	47N
Continuous current	1.4 A RMS	0.8 A RMS
Peak force	220N	140N
Forcer mass	1.3 kg	0.39 kg
Force constant	62.4 N/A-RMS	61.5 N/A-RMS
Magnetic attraction	809N	0N

As shown, the iron-core LSM requires more current than the ironless LSM, but is capable of delivering more force. Because of the iron in the forcer's windings, its forcer is over three times as heavy as that of the ironless motor.

The magnetic attraction is particularly interesting. In the iron-core motor, the attractive force is over 10 times the force delivered during regular operation. Even with skewed magnets on the track, it's safe to assume that cogging may be a problem for the SGLFW-35A.

8.2.3 Case Study: Transrapid Maglev System

The term *Transrapid* refers to three distinct entities:

- A set of maglev (magnetic levitation) trains built between 1970 and 2007.

- The company that constructed the Transrapid trains (called Transrapid International—a joint venture of ThyssenKrupp AG and Siemens AG).

- The design technology used to build the Transrapid trains, which refers back to a German patent in 1934.

Transrapid International has constructed nine train lines, and the most recent is the Transrapid SMT (Shanghai Maglev Train). This has many advantages over conventional trains, including greater speed, silent operation, and a lack of moving parts. It's also the fastest commercial train in current operation, carrying up to 574 passengers at speeds up to 268 miles per hour.

All of the Transrapid trains rely on iron-core linear synchronous motors for propulsion. Unlike the other LSMs discussed so far, the windings are located on the rail, which is commonly referred to as the *guideway* in maglev systems. Magnets are located on the underside of the train, and Figure 8.11 shows what the overall design looks like.

Figure 8.11
The Transrapid maglev
system

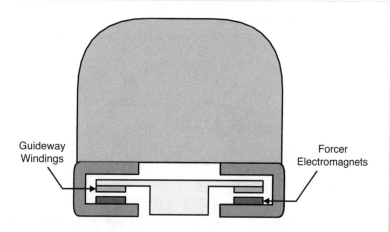

Guideway
Windings

Forcer
Electromagnets

As shown in the figure, the train wraps around the bottom of the guideway, where the guideway's windings interact with the train's magnets. These windings receive three-phase AC power, and the only portion of the guideway that receives power is the region directly under the train. The train can be brought to a halt by reversing the current in the windings.

Beneath the guideway's windings, the forcer magnets are DC-powered electromagnets, not permanent magnets. This is necessary because these magnets serve double-duty. In addition to their role in propulsion, they attract the train to the guideway, thereby making levitation possible. The power to these electromagnets is precisely controlled to keep the train at a distance of about 1 cm from the guideway.

8.3 Linear Induction Motors

Linear induction motors, or LIMs, have a great deal in common with LSMs and can be used in the same applications. Structurally speaking, the main difference is that LIMs don't have permanent magnets. Instead, motion is made possible by a conductor that receives induced voltage from the traveling magnetic field. The first part of this section discusses how LIMs work.

Like LSMs, LIMs play a central role in the field of maglev trains. The second part of this section presents Japan's LINIMO train line, which relies on a large-scale LIM for transport.

8.3.1 Structure and Operation

Chapter 6 introduced the topic of induction motors, which have two fundamental properties:

- AC power enters windings in the stator, which combine to produce a rotating magnetic field.

- The stator's changing field induces a voltage and current in the rotor's conductors, which results in a force that turns the rotor.

A linear induction motor is constructed by unrolling the elements of a rotary induction motor. That is, the stator of a linear motor consists of windings arranged in a line. Similarly, the rotor consists of a linear conductor facing the stator. Figure 8.12 shows what this looks like.

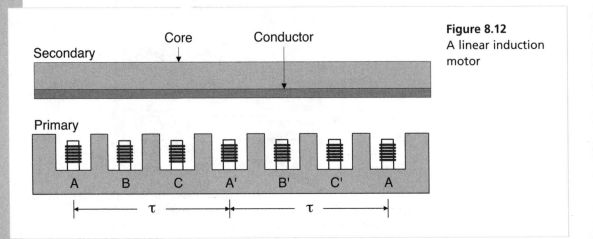

Figure 8.12
A linear induction motor

In this figure, the positions of the two elements can be reversed so that the windings move on top of the conductor. Because both elements can move, LIMs don't use the terms *rotor* and *stator*. Instead, the element that creates the changing magnetic field is called the *primary* and the element that receives induced voltage is called the *secondary*.

Linear induction motors are polyphase AC motors. In this case, the secondary receives three-phase power and directs the current to its A/A', B/B', and C/C' windings.

The changing field in the primary induces a voltage in the secondary's conductor. Voltage produces a current, and when a current-carrying conductor enters a changing magnetic field, the result is force.

A LIM's speed, denoted as v, is always less than its linear synchronous speed. The relationship between the two is called the slip, which can be computed with the following equation:

$$s = \frac{v_s - v}{v_s}$$

This is the same equation for slip as presented in Chapter 6. Note that slip is commonly expressed as a percentage.

8.3.2 The LINIMO Train Line

At the 2005 World Expo, Japan demonstrated one of the first commercially available maglev train lines: the Linear Motor line, or the LINIMO. This relies on a linear induction motor to transport passengers from Fujigaoka to Yagusa.

In 2004, researchers at the Tokyo University of Science released a report that provides information about the LINIMO's construction and operation. One notable point is that the motor makes use of a short stator design. That is, the primary is part of the train, and is positioned above the secondary. This differs from the long stator design employed by the Transrapid discussed earlier.

Figure 8.13 depicts a cross-section of the LINIMO train and platform. As shown, the overall structure is similar to that of the Transrapid.

Figure 8.13
Cross-section of the LINIMO system

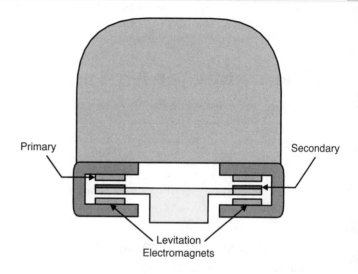

Relying on induced voltage in the secondary, the LINIMO commonly travels around 60 miles per hour. This is significantly slower than the Transrapid, which travels around 268 miles per hour. Also, the LINIMO moves with less force. The train line must be shut down when wind speeds exceed 25 m/s.

The primary receives three-phase AC power and distributes it to a series of windings. Figure 8.14 shows how a portion of these windings are arranged on one side of the primary.

Figure 8.14
Windings of the LINIMO primary

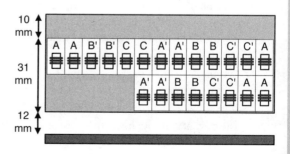

Each winding in the primary has three turns of wire and each receives 300 A. They're stacked into two tiers, with the windings of the second tier staggered relative to those in the first. As shown in the figure, the windings are only tens of millimeters in size. Personally, I find it incredible that such small electromagnets can move a large train.

The LINIMO's secondary is simpler in structure. A layer of aluminum is mounted to the top of an iron extension to the platform. This conductive layer is referred to as the *reaction plate*.

Below the primary, the train has electromagnets that keep the train at the proper distance from the platform. These are called levitation magnets or *lift-guide magnets* (LGMs). These electromagnets receive DC power, and as more current is applied, the attractive force between the magnets and the platform increases. A proximity sensor is used to ensure that the gap is kept to 8 mm.

8.4 Homopolar Motors

Homopolar motors are the oldest of the electrical motors, and the first was demonstrated in 1821. Like many of the motors discussed in this chapter, they can be rotary or linear. I would have mentioned them earlier in this book, but I've never seen a practical application of a rotary homopolar motor.

In contrast, linear homopolar motors have one famous (infamous?) application: the *railgun*. This section discusses this interesting topic, but first, it's important to understand the principle behind homopolar motors.

8.4.1 Structure and Operation

As discussed way back in Chapter 1, "Introduction to Electric Motors," when current-carrying conductors come near a magnet, the result is physical force. In every motor we've seen so far, the conductors and the magnet are positioned close to one another but never come into contact.

But what if the magnet becomes part of the circuit loop? Magnets are conductors. What if the current-carrying conductor and the magnet were connected? In this case, the resulting force on the conductor will be the same. However, there's one important difference: If the conductor can spin around the magnet, it will do so without the need of a brush. Therefore, homopolar motors can be thought of as brushless DC motors, but they're nothing like the BLDCs discussed in Chapter 3, "DC Motors."

The best way to understand how homopolar motors work is to see a demonstration, the most common of which involves a battery, a permanent magnet, and conductive wire. This has been performed countless times in classrooms, going back to the nineteenth century. If you search for homopolar motor demonstrations on the Web, you'll probably find a video whose setup looks like that in Figure 8.15.

Current flows from the battery's positive terminal to its negative terminal through the wires. Because the current-carrying wires are near the magnet's field, a force is produced that rotates them around the magnet. It's important to note that this demonstration drains the battery quickly.

The current's direction and the magnet's polarity (direction from north pole to south pole) remain the same throughout the motor's operation. This is why the motor is called homopolar. (*Homos* means "same" in Greek.)

Figure 8.15
Homopolar motor
demonstration

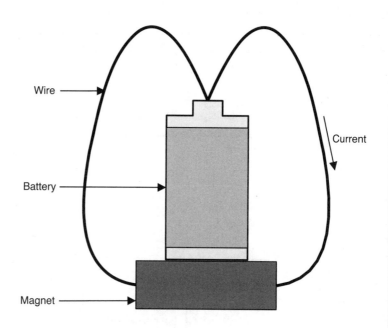

Wire

Current

Battery

Magnet

8.4.2 Railguns

To generate strong forces, motors rely on strong permanent magnets (such as the rare-earth magnets) or strong electromagnets, which are made up of coils of wire surrounding an iron core. However, every current-carrying conductor exhibits a degree of magnetic behavior. As more current flows through the conductor, this magnetic behavior grows stronger.

For example, consider the three conductors shown in Figure 8.16. The two long conductors are fixed in place and the short conductor is free to move between them. Each carries the same amount of current, denoted as I.

For normal levels of current, the magnetic behavior of the conductors is negligible. However, if the current grows large enough, up around 1000 A to 10,000 A, the magnetic behavior becomes strong enough to create a force that pushes the conductors apart.

The two long conductors won't move because they're fixed in place, but the force on the short conductor will push it to the right. This is shown in Figure 8.17.

The force increases with the current, and if the current is sufficiently large, the force can launch the small conductor as a projectile. In this case, the fixed conductors are referred to as *rails* and the structure is called a *railgun*.

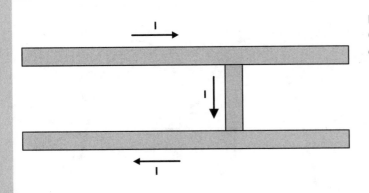

Figure 8.16
Current traveling through parallel conductors

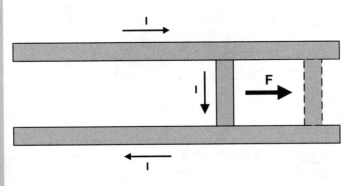

Figure 8.17
Railgun operation

The speed of a conventional bullet is limited by the chemical properties of gunpowder, but the speed of a railgun's projectile is limited only by the amount of current. For this reason, the U.S. Navy has experimented with railguns to launch projectiles from ships. In 2008, the Naval Surface Warfare Center demonstrated a railgun capable of launching a seven-pound conductive bullet at seven times the speed of sound.

The force exerted on the projectile is also exerted on the rails. Therefore, firing a railgun can cause significant damage to the gun itself. This damage and the extraordinary power requirements are the two main drawbacks of railgun technology.

It's important to see the differences between railguns and the coilguns discussed earlier in this chapter. A coilgun propels a ferromagnetic sabot through a coil of wire. The sabot doesn't make contact with the coil, so there's no damage to the system. However, the sabot's speed is limited by the magnetic properties of the coil. In contrast, the conductive projectile of a railgun is only limited by the amount of current delivered to the rails.

8.5 Summary

Many engineers think of linear motors as strange and foreign, with different terminology and operating principles. However, if you think of a linear motor as an unwrapped rotary motor, you shouldn't find the topic difficult to understand. Linear motors accept the same kind of power as rotary motors, and synchronous speed is computed in essentially the same way.

One point of confusion involves the terms *linear actuator* and *linear motor*. A motor is a device that produces motion, and an actuator is a type of motor intended to control another mechanism. The operation of an electric linear actuator is made possible by the tendency of an iron core (the plunger) to be pulled into an energized coil. A linear actuator that launches the plunger out of the coil is called a *coilgun*.

Rotary synchronous motors have a number of different types, but the vast majority of linear synchronous motors, or LSMs, are permanent magnet LSMs. In LSM terminology, the moving element is called the forcer, and the stationary element is called the rail or track. In an iron-core LSM, the forcer's windings have iron cores that increase force but also increase cogging. In an ironless LSM, the lack of iron makes for smoother motion, but reduces the amount of force. A slotless LSM combines aspects of iron-core and ironless motors, but has less efficiency than either.

Linear induction motors, or LIMs, are similar to LSMs, but have conductors in place of permanent magnets. These conductors receive induced current, which produces force in the presence of the windings' magnetic field. As with rotary induction motors, LIMs operate at a speed less than that of the motor's synchronous speed. The relationship between a LIM's speed and its synchronous speed is called slip.

The last section of this chapter discussed homopolar motors. These motors get their name from the fact that the electric current and magnetic field keep the same direction throughout the motor's operation. These motors are the oldest known electric motors, but they're also the least used. When it comes to homopolar motors, the only nontrivial application I know of is the railgun. This consists of two long conductors (the rails) that exhibit magnetic behavior when driven with high current. If the current is large enough, it can propel a third conductor with tremendous force.

This chapter has presented two maglev train lines: the Transrapid, which is based on LSM technology, and the LINIMO, which is based on LIM technology. Maglev trains have many advantages over regular trains, and after so many decades of research and development, you may wonder why we're not all riding them. The answer is cost. Maglev train lines are so expensive to construct that most governments don't consider them worth the investment. It's heartbreaking to see a fascinating technology fall into disuse, but until we see major developments in linear motors, it looks as though widespread maglev transport will remain out of reach.

9

MOTOR CONTROL WITH THE ARDUINO MEGA

The Arduino family of circuit boards is one of the core technologies that
have made the Maker Movement possible. If you're new to gadget development and you want to get your feet wet, Arduino boards are ideal
because they're so easy to use and easy to program. If you're an entre-
preneur who wants to manufacture and sell a gadget, Arduino boards are
ideal because they provide high reliability at low cost.

Many hundreds of thousands of Arduino boards have been sold, and
across the world, electronic hobbyists have incorporated them into
projects. Some projects are simple hobbyist gadgets, such as remote-
controlled musical instruments, but many others have become viable
products, including vehicles, household robots, and health-monitoring
systems.

The goal of this chapter is to explain how Arduino technology can be
used to control electric motors. There are three main parts:

- **The Arduino Mega**—Understanding the hardware and developing
software

- **The Arduino Motor Shield**—Understanding motor-control devices

- **Motor control**—Developing software to control a brushed motor,
stepper motor, and servomotor

This chapter covers a great deal of ground, and as much as I'd like to, I
can't explore any specific subject in detail. Thankfully, there are countless Arduino resources on the Internet. In particular, I recommend the

documentation page at http://arduino.cc/en/Reference/HomePage and the Arduino forum at http://forum.arduino.cc.

9.1 The Arduino Mega

The Arduino Mega isn't the most powerful or the most recent Arduino board, but it's one of the most popular. It's also fully compatible with existing Arduino hardware and software. Table 9.1 presents basic information about the board.

Table 9.1 Specifications of the Arduino Mega

Parameter Name	Parameter Value
Dimensions	4 × 2.1 inches
Operating voltage	5 V
Recommended input voltage	7–12 V
Clock speed	16 MHz
Digital I/O pins	54
Analog input pins	16

With its limited resources, the Mega isn't suitable for computing tasks such as text editing or web surfing. However, when combined with the Arduino Motor Shield, it's capable of controlling brushed motors, stepper motors, and servomotors. This section discusses the Arduino Mega's circuit board and its brain, the ATmega2560 microcontroller. The motor shield is introduced in a later section.

9.1.1 The Arduino Mega Circuit Board

The design of the Mega makes evident Arduino's focus on simplicity. The power pins are grouped together and labeled POWER. The communication pins are grouped together under the heading COMMUNICATION. Figure 9.1 shows what this looks like.

Most of the board's perimeter is occupied with raised black components that receive single-wire connections. These are called *headers*, and the Mega's header connections are divided into five groups:

- **Power**—Receive or deliver external power
- **Analog input**—Receive analog data to be converted into digital signals for processing
- **Digital**—Receive or transmit digital signals
- **Communication**—Communicate signals through three serial ports
- **PWM**—Transmit control signals using pulse width modulation

Figure 9.1
The Arduino
Mega

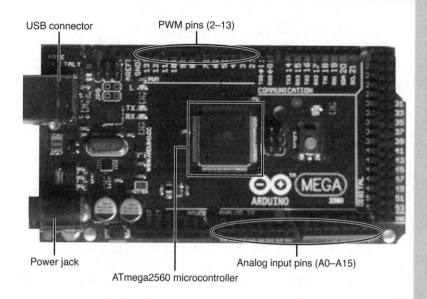

USB connector

PWM pins (2–13)

Power jack

ATmega2560 microcontroller

Analog input pins (A0–A15)

For the purposes of this chapter, the most important header signals are in the PWM group. We'll use these to generate control signals for DC motors.

The left side of the board has a power jack, and the recommended input voltage is between 7 and 12 volts. However, I prefer to deliver power to the Mega through the USB connector. If the Mega is connected to a PC by USB, it will draw the current it needs to operate.

In addition to delivering power, the USB connection also makes it possible to transfer programs to the Mega. However, before we get into Arduino programming, I want to introduce the device that executes the programs: Atmel's ATmega2560 microcontroller.

9.1.2 Microcontrollers and the ATmega2560

As shown in Figure 9.1, the center of the circuit board is occupied by a 100-pin device called the ATmega2560. This device is a *microcontroller*, and in essence the entire purpose of an Arduino board is to provide access to this device. This discussion explains what microcontrollers are and presents the specific characteristics of the ATmega2560.

Microcontrollers

In my experience, the best way to introduce microcontrollers is to compare them with personal computers. A PC serves a wide variety of purposes involving data processing and digital communication. To serve these purposes, the PC requires multiple devices—the CPU to process data, RAM chips to store temporary data, and the hard disk to store programs, files, and the operating system.

A microcontroller (abbreviated MCU) serves similar purposes, but all of its resources are integrated onto a single chip. This self-containment provides a number of benefits, including low cost, low power operation, and ease of circuit design. The drawback is that the MCU's on-chip resources aren't nearly as impressive as those you'd find in a PC.

An example will make this clear. My laptop has 8 gigabytes of RAM and its processor runs at 3 gigahertz. In contrast, the Mega's microcontroller has 8 kilobytes of RAM and processes data at 16 megahertz. This means my laptop runs nearly 200 times faster than the Mega and can store one million times as much data.

MCUs may not be suitable for personal computing, but they're ideal for maker projects. If you're building a simple robot or automated sensor system, the microcontroller's lack of resources won't be an issue. Instead, you'll appreciate the low cost, and because MCUs are so self-contained, you'll find it easy to design the circuit.

In general, gadgets with microcontrollers operate in three steps:

1. Read data from a sensor, such as a temperature sensor or a pressure sensor.

2. Process the data to judge the state of the system.

3. Use the state information to control a mechanism, such as a motor.

In Step 1, sensor data can take an infinite number of values, and for this reason, the data is referred to as *analog*. Before the data can be processed, it must be converted into the ones and zeros required by processors. This is *digital* data. To make this possible, most modern MCUs have multiple analog-to-digital converters, or ADCs.

In Step 3, microcontrollers control mechanisms using pulse width modulation (PWM), which has been discussed at length throughout this book. Later in this chapter, I'll show how the Mega can use PWM to control brushed motors and servomotors.

The ATmega2560

Most Arduino boards contain Atmel MCUs, and the Mega is no exception. The Mega relies on Atmel's ATmega2560 microcontroller to process its data, and Table 9.2 lists a number of its important characteristics.

Table 9.2 Characteristics of Atmel's ATmega2560 Microcontroller

Parameter Name	Parameter Value
Clock speed	16 MHz
Flash memory	256 KB
SRAM	8 KB
EEPROM	4 KB
Number of pins	100
Analog conversion resolution	10-bit
PWM resolution	8-bit
Temperature range	−40° to 85° C

In looking at these values, it's important to understand the difference between the three different types of memory:

- Flash memory holds programs, which means the largest program that can run on the Mega is 256 KB.

- SRAM (static random access memory) stores temporary data used by the program.

- EEPROM (electrically erasable programmable read-only memory) stores settings and other parameters.

The SRAM memory is cleared whenever power is removed. In contrast, the Flash memory and EEPROM maintain their contents without power.

The ATmega2560 MCU has 100 pins: 11 power pins and 89 pins for input/output. Most of the I/O pins can serve multiple roles, and configuring the pins' roles is a major concern in MCU development.

Thankfully, the Arduino framework makes pin configuration easy: 54 of the ATmega2560's I/O pins are accessible through the Mega's headers, and the process of configuring their operation is almost trivially simple.

The ATmega2560 has many incredible characteristics that have endeared it to makers across the world, but to really appreciate this device, you need to get your hands dirty and start writing programs. The next section explains how this is done.

9.2 Programming the Arduino Mega

In general, microcontroller programming isn't pleasant. You need to be aware of memory maps, peripheral buses, interrupt vectors, and countless data/control registers. Also, when you change to a new MCU, you practically have to rewrite your code from scratch.

The great innovation of the Arduino framework is that it dramatically simplifies writing code for MCUs. If you're familiar with the C programming language, you can come up to speed with Arduino in minutes—and as new Arduino boards come out, you can compile and run your programs without modification.

This section explains how to write, compile, and execute Arduino programs, commonly called *sketches*. However, before you can start programming, you need to get the Arduino environment up and running.

> **note**
>
> This chapter assumes a basic familiarity with C programming. If you don't have this background, my favorite introductory book is *C for Dummies* by Dan Gookin.

9.2.1 Preparing the Arduino Environment

The Arduino environment is a software package that consists of three components:

- USB drivers needed to communicate with Arduino boards

- A compiler that converts sketch code into executables for microcontrollers

- An integrated development environment (IDE) for writing, editing, compiling, and uploading sketches

The Arduino environment can be downloaded from http://arduino.cc/en/Main/Software. Two releases are available, and it's important to understand the difference between them:

- **Arduino 1.0.x**—This environment is stable and supports boards with 8-bit processors such as the Arduino Mega.

- **Arduino 1.5.x**—This environment supports boards with 32-bit microcontrollers, such as the Arduino Yún and the Arduino Due. At the time of this writing, this is a *beta release*, and to quote the site: "You may encounter bugs or unexpected behaviours."

To program the Arduino Mega, you'll need the first option. To download it, open http://arduino.cc/en/Main/Software in a browser, scroll until you see the Arduino 1.0.x heading, and click the link corresponding to your operating system.

After you've downloaded the Arduino file, you're ready to install the application. The installation process depends on your operating system:

 note

The newer environment software (version 1.5.x) has been in beta for a long time. Many people are happy with it, but I've found it to be unreliable. This is why this chapter relies on the Arduino Mega for motor control instead of a newer board.

- For Windows, the instructions are at http://arduino.cc/en/Guide/Windows.

- For Mac OS X, the instructions are at http://arduino.cc/en/Guide/MacOSX.

- For Linux, the operating instructions depend on the distribution. The list of supported distributions and their instructions can be found at http://playground.arduino.cc/Learning/Linux.

If you've completed the directions, you should have an executable that brings up the Arduino IDE. Figure 9.2 shows what this looks like on my Windows 7 computer.

When the environment starts for the first time, it has no idea what Arduino board you're using or how to access it. Therefore, before you start coding, you need to tell the environment about your board. For boards connected by USB, I recommend five steps:

1. Connect the Arduino Mega to a USB cable and connect the other end of the cable to a PC.

2. Launch the Arduino environment.

3. In the main menu, go to Tools > Board and select the option Arduino Mega 2560 or Mega ADK.

4. In the main menu, go to Tools > Serial Port and select the serial port to which the Arduino Mega is connected. The process of determining the connected serial port depends on the operating system.

5. In the main menu, go to File > Save As and enter **blink** for the name of the sketch. Click Save.

After these steps are completed, the environment should look similar to that shown in Figure 9.3.

Figure 9.2
The Arduino development
environment

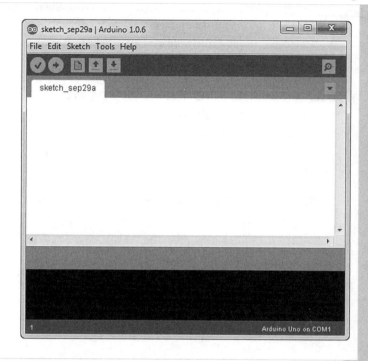

Figure 9.3
The fully configured environment

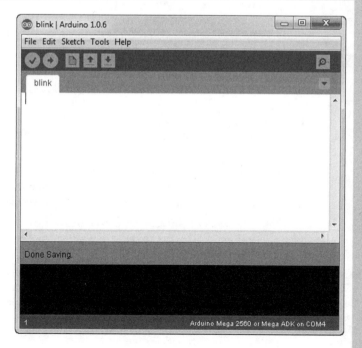

The final step saves an empty sketch to a file called blink.ino (*.ino is the extension for Arduino sketches). By default, this is saved in the Arduino\blink directory, which is located in the user's documents folder. For example, on my Windows system, blink.ino is saved to the C:\Users\Matt\My Documents\Arduino\blink directory.

9.2.2 Using the Environment

Once you've configured the environment for your board, the hard part is over. Now all you need to do is edit code and use the buttons above the editor.

To explain how this is done, I'm going to walk through the process of compiling and uploading a simple sketch. Listing 9.1 presents the code for blink.ino.

Listing 9.1   Ch9/blink.ino—Causing an LED to Blink

```
/* This sketch sets the voltage of Pin 13 high and low.
This causes the LED connected to the pin to blink. */

// Assign a name to Pin 13
int led_pin = 13;

// At startup, configure Pin 13 to serve as output
void setup() {
  pinMode(led_pin, OUTPUT);
}

// Repeatedly change the voltage of Pin 13
void loop() {
  digitalWrite(led_pin, HIGH);  // set the pin voltage high
  delay(1000);                  // delay one second
  digitalWrite(led_pin, LOW);   // set the pin voltage low
  delay(1000);                  // delay one second
}
```

If you'd rather not code this by hand, you can load my blink.ino sketch. Every code example in this book is contained in an archive called mfm.zip, which can be freely downloaded from http://www.motorsformakers.com.

After downloading the archive and extracting its contents, you'll find blink.ino in the Ch9 directory. You can load this into the environment by going to the main menu and selecting File > Open.

After the code has been entered, the next step is to compile it into a binary suitable for the Mega. This is accomplished by clicking the leftmost button above the editor, which features a check mark. If the code has errors, an error message in orange text will identify the first error and the line of code that produced it. If there are no errors, a "Done compiling" message will be displayed.

 note

If you enter the code by hand, you'll see the number in the lower-left corner of the editor change as you type. This identifies the line number the cursor is on, and it took me far too long to figure that out.

If clicked, the button with a right-pointing arrow will recompile the sketch and upload it to the board. If this succeeds, a "Done uploading" message will be displayed and the program will start executing on the board. On the Arduino Mega, the LED next to Pin 13 will start blinking—one second on, one second off.

9.2.3 Arduino Programming

An Arduino program consists of statements that have the same structure and syntax as statements in the C programming language. That is, each statement ends with a semicolon and statements can be grouped into named blocks called *functions*. Arduino supports many of the basic C data types, and for many Mega programs the only data type you'll need is the `int` type.

Unlike C, sketches don't have a top-level `main` function. Instead, every sketch can be divided into three parts:

- **Global variables**—This part declares and initializes variables that can be used throughout the sketch.

- `setup()`—Contains statements to be executed when the board starts up or resets.

- `loop()`—Contains statements to be repeated after the `setup` function finishes.

A simple example will clarify how Arduino programs work. The code in Listing 9.1 repeatedly sets the voltage for Pin 13 high and low, delaying one second with each change. This causes the LED connected to Pin 13 to blink.

To understand the blink sketch, you need to be familiar with the functions provided by the Arduino framework. This discussion doesn't cover all of them, but focuses on functions in four categories: digital I/O, timing, analog read, and analog write. Table 9.3 lists each of them with a description.

 note

`HIGH` and `LOW` are regular `int` values. `HIGH` equals 1 and `LOW` equals 0.

Table 9.3 Important Sketch Functions

Category	Function	Description
Digital I/O	`pinMode(int pin_num,` `        int mode_type)`	Configures whether a pin's mode is `INPUT`, `OUTPUT`, or `INPUT_PULLUP`
	`digitalRead(int pin_num)`	Returns `HIGH` or `LOW`, depending on the input voltage

	`digitalWrite`
	`(int pin_num,`
	`int level)`
		Sets the output pin's voltage to `HIGH` or `LOW`

Timing	`delay(int time)`	Waits a number of milliseconds before completing
	`delayMicroseconds` `(int time)`	Waits a number of microseconds before completing

Category	Function	Description
	`millis()`	Returns the number of milliseconds since the program started
	`micros()`	Returns the number of microseconds since the program started
Analog read	`analogReference` `(int ref_type)`	Sets the maximum voltage for analog input
	`analogRead()`	Returns the input analog voltage

Analog write	`analogWrite`
	`(int pin_num,`
	`int duty_cycle)`
		Delivers PWM pulses with the desired duty cycle

This table lists less than half of the functions available for Arduino programming. For the full list, visit the reference site at http://arduino.cc/en/Reference/HomePage.

Digital I/O

As discussed earlier, the Mega's pins can be divided into five groups: power, analog input, communication, digital, and PWM. With the exception of the power pins, every pin on the Mega can be configured to serve one of three roles:

- INPUT—An input pin's digital voltage level (HIGH or LOW) can be read with `digitalRead`.

- INPUT_PULLUP—An input pin whose default state is HIGH.

- OUTPUT—An output pin's voltage can be set with `digitalWrite`.

A pin's mode determines whether it's an input pin or an output pin. To configure this mode, the `pinMode` function requires two arguments: the pin's number and the desired mode. By default, every pin's mode is set to INPUT. The following code sets Pin 10 to behave as an OUTPUT pin:

```
pinMode(10, OUTPUT);
```

If `pinMode` sets a pin in INPUT or INPUT_PULLUP mode, `digitalRead` returns an int corresponding to its voltage level. For INPUT mode, `digitalRead` returns HIGH if connected to a voltage greater than 3 V and LOW if connected to a voltage less than 2 V. For INPUT_PULLUP mode, `digitalRead` returns HIGH by default, and returns LOW when the pin is connected to ground.

If `pinMode` sets a pin's mode to OUTPUT, then `digitalWrite` can be called to set its voltage level. This function accepts two arguments: the pin number and the voltage level. If the second argument is HIGH, then `digitalWrite` sets the pin's voltage to 5 V. If the second argument is low, the pin's voltage is set to 0 V.

As a brief example, the following code reads the voltage level of Pin 7 and writes the voltage level to Pin 9:

```
res = digitalRead(7);
digitalWrite(8, res);
```

If this code is placed inside the `loop` function, it will be executed repeatedly. If it's placed inside the `setup` function, it will be run once per execution of the program.

Timing

The Arduino timing functions are easy to understand and use. There are four in total: Two specify timing delays and two identify how long the program has been running.

When a program updates a pin's state, such as when `digitalWrite` changes a pin's voltage, you may want to maintain this state for a duration of time. This is made possible by the `delay` and `delayMicroseconds` functions.

For example, the blink application sets Pin 13's voltage from `HIGH` to `LOW` and back again. With each change, the `delay` function maintains the state for 1 second. This is accomplished with the following code:

```
delay(1000);
```

Once `delay` starts, further statements won't execute until it finishes. Its argument identifies the wait time in milliseconds. If the argument is 250, `delay` will prevent statements from executing for a quarter of a second.

In many applications, a millisecond can be too long. If this is the case, you can call `delayMicro-seconds`. This is similar to `delay`, but the argument identifies the wait time in microseconds. A microsecond is one-thousandth of a millisecond, so `delayMicroseconds(500)` waits for 500 micro-seconds, which equals one-half of a millisecond, which equals 0.0005 seconds.

In a sketch, the `loop` function executes until power is removed. This means you can't halt the `loop` function or break out of it. However, if you want to execute code for a specific duration of time, you can use the `millis` and `micros` functions. These tell you how long the program has been running, and their return values are given in milliseconds and microseconds, respectively.

An example will show how `millis` is used in practice. The following code sets Pin 13 to `HIGH` for the first 5 seconds, `LOW` for the next 5 seconds, and back to `HIGH`:

```
if (millis() < 5000)
  digitalWrite(13, HIGH);
else if (millis() < 10000)
  digitalWrite(13, LOW);
else
  digitalWrite(13, HIGH);
```

`micros` allows more precise time measurement. This can be helpful when you're controlling a mechanism or communicating with another device.

Analog Read

If you want to read data from a sensor or another analog device, Arduino provides two crucial functions: `analogReference` and `analogRead`.

An analog signal can take an infinite number of values, but the Mega's pins can't read an infinite number of values. Therefore, when writing a sketch that reads analog data, you need to know the maximum voltage that can be read.

By default, the maximum analog voltage that can be read by the Mega is 5 V. This means the analog inputs can only distinguish inputs between 0 V and 5 V. This maximum can be changed with the `analogReference` function, whose argument can take one of four values:

- `DEFAULT`—The default value of 5 V.

- `INTERNAL1V1`—A maximum of 1.1 V.

- `INTERNAL2V56`—A maximum of 2.56 V.

- `EXTERNAL`—The maximum is set by the voltage on the AREF pin.

If the board is oriented as shown in Figure 9.1, AREF is the leftmost pin in the top header. If `analogReference` is called with its argument set to `EXTERNAL`, the board's analog pins will read input voltages between 0 V and the voltage on AREF. Note that AREF must be set to a voltage between 0 V and 5 V.

Like `digitalRead`, `analogRead` accepts the pin number that the voltage should be read from. Also like `digitalRead`, `analogRead` returns an `int`. However, there are two major differences between these functions:

- The `int` returned by `analogRead` ranges from 0 and 1023, where 0 represents a voltage of 0 V and 1023 represents the maximum voltage.

- `analogRead` can only be called for specially configured analog input pins. As shown in Figure 9.1, the Arduino Mega has 16 analog input pins, A0–A15.

An example will make this clear. The following code reads the analog voltage on Pin A5:

```
analog_v = analogRead(A5);
```

Another point about the analog input pins is that they can be accessed by the digital I/O functions, `digitalRead` and `digitalWrite`. For example, the following code reads the digital voltage level from Pin A5:

```
digital_v = digitalRead(A5);
```

Like regular digital pins, the analog input pins are configured in `INPUT` mode by default. With the `pinMode` function, they can be configured in the `OUTPUT` or `INPUT_PULLUP` mode.

Analog Write

The `analogWrite` function is so important that it deserves its own category. When I first saw this function, I assumed the microcontroller had digital-to-analog converters (DACs) capable of converting integer values into true analog outputs. Unfortunately, the Mega doesn't have any DACs, so it's incapable of producing real analog values.

Instead, `analogWrite` on the Mega produces a train of pulses formatted with pulse width modulation (PWM). As introduced in Chapter 2, "Preliminary Concepts," PWM is the primary mechanism for controlling most DC motors. Pulses in a PWM signal have the same height and period, but the pulse width may vary over the course of the signal. The ratio of the pulse width to the period is the duty cycle.

To generate a PWM signal, the `analogWrite` function needs two arguments:

- **Pin number**—`analogWrite` is available only for a specific set of pins (2–13, 44–46 on the Mega). It cannot be called on the analog input pins.

- **Duty cycle**—This value determines the time width of the pulse relative to the time between pulses. This takes a value between 0 (always off) and 255 (always on).

The code in Listing 9.2 shows how `analogWrite` can be used to generate pulses.

Listing 9.2 Ch9/pwm.ino—Pulse Width Modulation

```
/* This sketch produces a pulse-width modulation (PWM) signal
whose duty-cycle switches between 0%, 25%, 50%, and 75%. */

// Assign a name to Pin 13
int pwm_pin = 13;

// Configure Pin 13 as an output pin
void setup() {
  pinMode(pwm_pin, OUTPUT);
}

// Switch the duty-cycle between 25% and 75%
void loop() {
  analogWrite(pwm_pin, 0);      // set duty cycle to 0%
  delay(1000);                  // delay one second
  analogWrite(pwm_pin, 64);     // set duty cycle to 25%
  delay(1000);                  // delay one second
  analogWrite(pwm_pin, 128);    // set duty cycle to 50%
  delay(1000);                  // delay one second
  analogWrite(pwm_pin, 192);    // set duty cycle to 75%
  delay(1000);                  // delay one second
}
```

In this sketch, `setup` configures Pin 13 in `OUTPUT` mode. Then `loop` calls `analogWrite` four times, changing the duty cycle from 0% to 25% to 50% to 75%. This changes the brightness of the LED connected to Pin 13. After each change, the sketch delays for one second. Figure 9.4 gives an idea of what these pulses look like.

The time between pulses (period) varies from board to board, and occasionally from pin to pin. Based on my tests on the Mega, the period for Pins 2, 3, and 5–12 is 2.05 ms. The period for Pins 4 and 13 is about 1.025 ms. Put another way, the PWM frequency for Pins 2, 3, and 5–12 is 488 Hz and the PWM frequency for Pins 4 and 13 is 976 Hz.

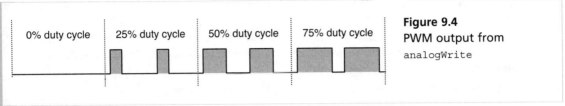

| 0% duty cycle | 25% duty cycle | 50% duty cycle | 75% duty cycle |

Figure 9.4
PWM output from
`analogWrite`

It's possible to change the PWM frequencies of the Arduino Mega with special code. This topic lies beyond the scope of this book, but there are plenty of online resources that explain how it can be done.

9.3 The Arduino Motor Shield

The Arduino Mega has many capabilities, but it can't deliver enough current to control a motor. It also lacks the H bridge needed to reverse a motor's direction. Therefore, before the Mega can control a motor, it must be connected to the Arduino Motor Shield.

In Arduino parlance, a *shield* is a secondary circuit board that can be connected on top of an Arduino board. Many different types of shields are available for Arduino devices, including shields for wireless communication, GPS tracking, and MP3 playing. The motor shield contains resources needed for motor control, and Figure 9.5 shows what it looks like.

The motor shield may seem confusing because of the many connections to support different types of motors. The goal of this section is to explain how it works. Later sections explain how the shield can be used to control specific motors.

9.3.1 Power

The logic devices on the motor shield receive power from the Arduino Mega, but this isn't sufficient to deliver power to a motor. For this reason, the motor shield has its own power connections: the Vin and GND screw terminals in the lower-left of the figure. Vin accepts voltages between 7 V and 12 V and can receive as much as 2 A of current per motor.

If Vin is set to a high voltage, it's important to keep the shield's power from affecting the Arduino Mega. The official directions recommend removing the "Vin Connect" jumper on the underside of the shield. I recommend bending the shield's Vin pin (at the far right of the POWER header) so that it doesn't connect to the Mega.

Above the Vin and GND screw terminals, the motor shield has four connections for output power. These can provide power to two brushed motors or one stepper motor. Later sections explain how this is accomplished.

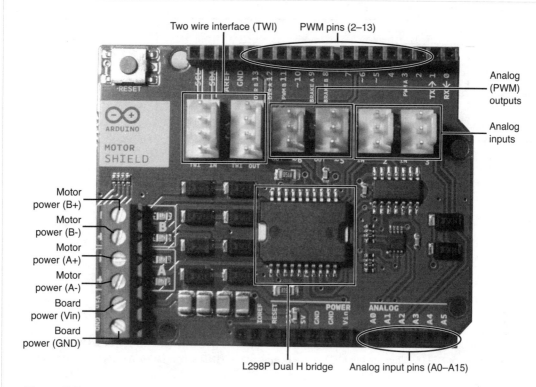

Two wire interface (TWI)   PWM pins (2–13)

Analog
(PWM)
outputs

Analog
inputs

Motor
power (B+)

Motor
power (B-)

Motor
power (A+)

Motor
power (A-)

Board
power (Vin)

Board
power (GND)

L298P Dual H bridge   Analog input pins (A0–A15)

Figure 9.5
The Arduino Motor Shield (v1.1)

9.3.2 The L298P Dual H Bridge Connections

Chapter 3, "DC Motors," introduced the H bridge, whose four switches make it possible to reverse current to a motor. The motor shield contains two H bridges in the form of the L298P integrated circuit. This chip uses bipolar junction transistors (BJTs) to serve as switches, and Figure 9.6 shows how the first H bridge is connected to the shield's signals.

This circuit is complex, but keep in mind that its primary purpose is to deliver power to the motor outputs, MOT_A+ and MOT_A-. To turn the motor in the forward direction, MOT_A+ should be connected to Vin and MOT_A- should be connected to GND. To reverse the motor's direction, MOT_A+ should be connected to GND and MOT_A- should be connected to Vin.

The PWM_A signal receives PWM pulses from the Arduino Mega board. When this is high, the circuit functions normally. When it's low, no voltage is applied to the switches' inputs, which means MOT_A+ and MOT_A- are left unconnected.

If PWM_A is high, the states of the four switches are controlled by DIR_A and BRAKE_A. If DIR_A is high, S_0 connects MOT_A+ to Vin. If DIR_A is low, S_2 connects MOT_A+ to GND.

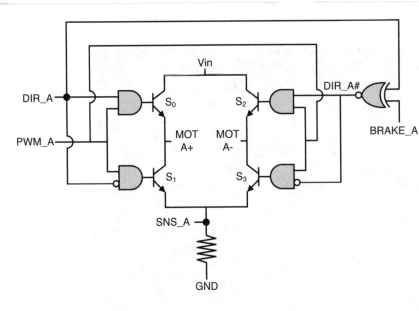

Figure 9.6
H bridge connections

The right of the diagram shows how BRAKE_A affects the circuit. When BRAKE_A is low, DIR_A# is the inverse of DIR_A. This means MOT_A- is connected to GND when MOT_A+ is connected to Vin, and vice versa.

However, when BRAKE_A is high, DIR_A# equals DIR_A. This means MOT_A+ and MOT_A- are always connected to the same source. Because the voltage difference between MOT_A+ and MOT_A- is zero, no current flows through the motor, so it comes to a halt.

To control a motor properly, it's important to know how the signals in Figure 9.6 relate to the Arduino Mega's pins. Table 9.4 lists each of the motor signals and their corresponding pins.

Table 9.4 Motor Signals and Arduino Pins

Motor Signal	Arduino Mega Pin	Description
DIR_A	12	Controls the direction of Motor A
DIR_B	13	Controls the direction of Motor B
PWM_A	3	PWM signal for Motor A
PWM_B	11	PWM signal for Motor B
BRAKE_A	9	Halts Motor A when high
BRAKE_B	8	Halts Motor B when high
SNS_A	A0	Current sensing for Motor A
SNS_B	A1	Current sensing for Motor B

9.3.3 Controlling a Brushed Motor

An example will clarify how the dual H bridge and its connections can be used to control a motor. The code in Listing 9.3 controls a brushed DC motor whose wires are connected to the shield's Motor Power A+ and Motor Power A- screw terminals.

Listing 9.3   Ch9/brushed.ino—Brushed DC Motor Control

```
/* This sketch controls a brushed motor. It drives it in the
forward direction at 75% duty cycle and halts. Then it
drives it in reverse at 75% duty cycle and halts. */

// Assign names to motor control pins
int dir_a = 12;
int pwm_a = 3;
int brake_a = 9;

// Configure the motor control pins in output mode
void setup() {
  pinMode(dir_a, OUTPUT);
  pinMode(pwm_a, OUTPUT);
  pinMode(brake_a, OUTPUT);
}

// Deliver power to the motor
void loop() {

  // Drive the motor forward at 75% duty cycle
  digitalWrite(brake_a, LOW);
  digitalWrite(dir_a, HIGH);
  analogWrite(pwm_a, 192);
  delay(2000);

  // Halt the motor for a second
  digitalWrite(brake_a, HIGH);
  delay(1000);

  // Drive the motor in reverse at 75% duty cycle
  digitalWrite(brake_a, LOW);
  digitalWrite(dir_a, LOW);
  analogWrite(pwm_a, 192);
  delay(2000);

  // Halt the motor for a second
  digitalWrite(brake_a, HIGH);
  delay(1000);
}
```

When the processing loop starts, DIR_A is set to 1 and PWM_A is set to 192. This drives the motor forward at 75% duty cycle. After a halt period, DIR_A is set to 0 and PWM_A is set to 192. This drives the motor in reverse at 75% duty cycle.

9.4 Stepper Motor Control

Of the many motors discussed in this book, stepper motors are the easiest to understand: They rotate through a fixed angle and halt. But they aren't the easiest motors to control. Bipolar steppers have four connections that require signals, and unipolar steppers have six.

The motor shield makes it straightforward to control a stepper. Not only is the shield's hardware ideally suited for the purpose, but Arduino provides free software to get your sketches working. This free software is packaged in the form of a *library*, so the first part of this section explains how to obtain the Stepper library and use its functions.

9.4.1 The Stepper Library

When you install the Arduino environment, you can call about 40 functions, not including the Stream and Serial functions. This set of functions can be extended using libraries. For example, one library contains functions for communicating across a serial peripheral interface (SPI) bus. Another contains functions that control liquid crystal displays (LCDs).

To see what libraries are available, visit http://arduino.cc/en/Reference/Libraries. Most of these libraries fall into one of two categories: standard libraries and contributed libraries. Contributed libraries must be downloaded and installed into the Arduino environment. Standard libraries don't have to be downloaded or installed. They're already included in the environment.

To access functions from a standard library, open a sketch in the environment's editor. Then, in the main menu, go to Sketch > Import Library and select the library you're interested in. This section relies on the Stepper library, and if you select this option, the following code will be prepended to the sketch:

```
#include <Stepper.h>
```

With this line added to the sketch, you can call the functions of the Stepper library. Table 9.5 lists each of them and provides a description of its purpose.

Table 9.5 Functions of the Stepper Library

Function	Description
Stepper(int steps_per_rev, int pin1, int pin2)	Returns a `Stepper` object with the given number of steps per revolution and connection pins
Stepper(int steps_per_rev, int pin1, int pin2, int pin3, int pin4)	Returns a `Stepper` object with the given number of steps per revolution and connection pins
setSpeed(int rpm)	Used to set the stepper speed in revolutions per minute
step(int steps)	Used to tell the stepper to turn one or more steps

These functions are straightforward if you know what objects and classes are. Just in case you don't, I'll provide a brief overview of object-oriented theory. Then I'll explain how to use the functions in Table 9.5.

Objects and Classes

The first two functions in Table 9.5 aren't like any of the other functions discussed in this chapter. That is, they're not called in the setup or loop method. Instead, their purpose is to create a new global variable, and for this reason, the Stepper function must be coded above the setup function.

The variable created by the Stepper functions isn't an int or a float, but has the Stepper data type. Technically speaking, Stepper is a *class*, and any variable created by the Stepper method is an *object*. Object-oriented programming (OOP) is a deep topic, and many books have been written on it. However, for Arduino development, you need to understand only four points:

- Every object is created by a function called a *constructor*. A class may have multiple constructors, and each must have the same name as its class.

- An object may contain variables of its own. These are called *member variables*.

- An object may contain functions of its own. These are called *member functions*.

- An object's member variables and functions can be accessed by following the object name with a dot (.) and the name of the variable or function.

In Table 9.5, the first two functions are constructors and the second two functions are member functions. As an example, the following code creates a Stepper object using the first constructor in the table. Later in the sketch, the loop function calls one of the object's member functions.

```
Stepper s = Stepper(200, 6, 5);
...
loop() {
  ...
  s.step(1);
  ...
}
```

The first line calls the Stepper constructor. Like a regular function, the constructor accepts arguments and returns a variable. In this case, the variable is a Stepper object named s.

This object has two member functions it can call: setSpeed and step. In this code, the object's step function is called inside loop. The dot between s and step identifies step as a member of the s object. It's important to note that a member function must be called with its associated object.

Stepper Functions

In both Stepper constructors, the first argument sets the number of steps the motor needs to turn to complete a revolution. For example, if each turn of the stepper rotates by an angle of 1.8°, the number of steps in a revolution is 360°/1.8° = 200. The constructor requires that this be given as an int.

The constructors' other arguments identify which pins control the stepper motor. Depending on the nature of the motor, it may be connected to two or four pins.

After the `Stepper` object is created, `setSpeed` specifies the desired speed of the motor. By combining the steps per revolution and the revolutions per minute, the program determines the time delay between steps.

For example, suppose the motor takes 150 steps to complete a revolution and `setSpeed` is called with a speed of 20 revolutions per minute. The program will determine that the motor should take 150 × 20 = 3000 steps in a minute, or 50 steps per second. Therefore, the program will delay 1/50 = 0.02 seconds between steps.

The last function in the table is `step`, whose argument identifies how many times the motor should step. If the argument is 1, the motor will step once and control will return to the program. If the argument is greater than 1, the motor will step multiple times. In this case, the program will halt until the steps are completed. If the argument is negative, the stepper will rotate in the reverse direction.

9.4.2 Controlling a Stepper Motor

As discussed in Chapter 4, "Stepper Motors," stepper motors come in two types: bipolar and unipolar. As a quick review, here are their characteristics:

- Bipolar steppers have four wires. Unipolar motors have five or six.

- Bipolar steppers require two H bridges for control. Unipolar stepper control is less complex.

- Bipolar steppers are significantly more efficient because they utilize the entire length of each winding when energized.

The drawback of using bipolar steppers is the need for two H bridges, but the motor shield's L298P provides two H bridges, so this isn't a concern. Therefore, this discussion focuses on controlling bipolar steppers. If you have a unipolar stepper, these directions still apply. Just ignore the center tap connections and connect the remaining wires as needed for a bipolar stepper.

Both types of stepper motors have two phases: A/A' and B/B'. Figure 9.7 shows how these phases are related to motor's windings and external connections.

Looking back at Figure 9.6, it should be clear that A and A' should be connected to the MOT_A+ and MOT_A- screw terminals in the lower-left corner of the shield. Similarly, B and B' should be connected to MOT_B+ and MOT_B-. In Figure 9.5, these connections are labeled Motor Power (A+), Motor Power (A-), Motor Power (B+), and Motor Power (B-).

The output of the two H bridges is determined by DIR_A and DIR_B, which correspond to Pins 12 and 13. Therefore, these are the pins provided to the `Stepper` constructor. This is shown in Listing 9.4.

 note

Don't be concerned if the colors in Figure 9.7 don't match the wires of your stepper. You can tell which wires are paired together by testing the resistance between them. Also, it doesn't matter if you connect B/B' in place of A/A'. The stepper will still rotate.

Figure 9.7
Connections of a bipolar
stepper

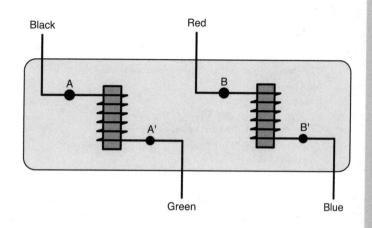

Listing 9.4 Ch9/stepper.ino—Stepper Motor Control

```
/*
This sketch controls a bipolar stepper motor,
stepping ten times in the forward direction and
ten times in the reverse direction.
The steps/revolution is set to 200 (1.8 deg/step)
and the speed is set to 10 RPM.
 */

#include <Stepper.h>

// Set the pin numbers
int pwm_a = 3;
int pwm_b = 11;
int dir_a = 12;
int dir_b = 13;

// Create a stepper object
Stepper s = Stepper(200, dir_a, dir_b);

void setup() {

  // Set speed to 10 revs/min
  s.setSpeed(10);

  // Make sure the two H Bridges are always on
  pinMode(pwm_a, OUTPUT);
  pinMode(pwm_b, OUTPUT);
  digitalWrite(pwm_a, HIGH);
```

```
    digitalWrite(pwm_b, HIGH);
}

void loop() {

    // Ten steps in the forward direction
    s.step(10);
    delay(1000);

    // Ten steps in the reverse direction
    s.step(-10);
    delay(1000);
}
```

In addition to DIR_A and DIR_B, this sketch sets the values of PWM_A and PWM_B. These signals need to be set high to ensure that both H bridges will function normally. Note that, when controlling a stepper, you don't really need PWM signals or braking.

9.5 Servomotor Control

Chapter 5, "Servomotors," discussed the topic of servos and how most servos used by makers don't provide feedback to the controller. This section explains how to control these hobbyist servos by accessing the Arduino Motor Shield. The first part of the section discusses the Servo library and its functions. The second part presents the servo-control code.

9.5.1 The Servo Library

Like the Stepper library discussed earlier, the Servo library is a standard library, which means it comes preinstalled with the Arduino environment. To access this library in a sketch, go to the main menu, select Sketch > Import Library, and select the Servo option. This will prepend the following code to the sketch:

```
#include <Servo.h>
```

Just as the Stepper library defines a class called `Stepper`, the Servo library defines a class called `Servo`. This class contains a number of methods, and Table 9.6 lists each of them.

Table 9.6 Functions of the Servo Library

Function	Description
attach(int pin)	Associates the Servo object with the given pin
attach(int pin, int min, int max)	Associates the Servo object with the pin and sets the pulse widths for the min/max servo angle
attached()	Returns 1 if the servo is attached to a pin and 0 if not

Function	Description
detach()	Used to delete the association between the Servo object and its pin
write(int angle)	Used to set the servo's angular position
writeMicroseconds(int time)	Used to set the pulse width of the signal delivered to the servo
read()	Returns the last angle written to the servo

None of these functions is a constructor, so these functions can't create `Servo` objects. Instead, a `Servo` object can be declared like any other variable. This is shown in the following line of code:

```
Servo sv;
```

The `Servo` object must be associated with the pin that will deliver PWM control signals to the motor. This association is created by the `attach` function, which can be called with one argument or three arguments.

When you're controlling a servo's shaft, the minimum pulse width produces the minimum angle (usually 0°) and the maximum pulse width produces the maximum angle (usually 180°). If `attach` is called with one argument (the PWM pin number), the minimum pulse width is 544 and the maximum pulse width is 2400.

If `attach` is called with three arguments, the first argument is the PWM pin number, the second is the minimum pulse width, and the third is the maximum pulse width. As an example, the following code associates `sv` with Pin 8 and sets the min/max pulse widths to 900/2100:

```
sv.attach(8, 900, 2100);
```

After the `Servo` object is associated with its pin, the angle of the motor's shaft can be set with `write` or `writeMicroseconds`. The `write` function accepts the desired angle in degrees, and the program determines the appropriate pulse width. If the desired pulse width is already known, `writeMicroseconds` accepts the pulse width in milliseconds.

9.5.2 Controlling the Servomotor

As discussed in Chapter 5, hobbyist servomotors generally have three wires:

- **Power**—Provides about 5–6 V to power the motor (usually the red wire)
- **Signal**—Controls the servo with PWM signals
- **Ground**—Provides electrical ground (usually the black wire)

In Figure 9.5, the shield has two orange connectors labeled Analog (PWM) Outputs. These connectors have three pins each: power, PWM control, and ground, in that order.

Unfortunately, the connections of my hobbyist servos are ordered with the PWM control signal first, followed by the power and ground. For this reason, I've found it necessary to separate the servomotor's wires and manually connect them to the shield's header connections.

The code in Listing 9.5 shows how to control a servomotor whose PWM control wire is connected to Pin 6.

Listing 9.5 Ch9/servo.ino—Servomotor Control

```
/*
This sketch controls a hobbyist servomotor.
It rotates the shaft 180 degrees forward and 180 degrees back.
 */

#include <Servo.h>

Servo sv;        // Servo object
int angle;       // servo's angular position

void setup() {

  // Attach the Servo object to Pin 6
  sv.attach(6, 800, 2200);
}

void loop() {

  // Rotate from 0 to 180 degrees
  for(angle = 0; angle < 180; angle += 1) {
    sv.write(angle);
    delay(10);
  }

  // Rotate from 180 to 0 degrees
  for(angle = 180; angle >= 1; angle -=1) {
    sv.write(angle);
    delay(10);
  }
}
```

In this case, the minimum angle corresponds to a pulse width of 800 microseconds and the maximum angle corresponds to a pulse width of 2200 microseconds. These values are specified in the `attach` function.

9.6 Summary

After so many chapters discussing motor theory, it's good to see how real-world motors can be controlled with real-world electronics. Many motor control systems cost a great deal of money, but the Arduino Mega and Arduino Motor Shield are both inexpensive and easy to use. With the Arduino programming environment, you can have a motor control sketch coded and compiled in less than an hour.

The primary device of the Arduino Mega is the ATmega2560 microcontroller. This single chip contains all the ROM, RAM, and processing power needed to execute Arduino sketches, but the Mega doesn't have the resources needed to control motors.

In contrast, the Arduino Motor Shield is designed to control brushed DC motors, stepper motors, and servomotors. Its primary device is the L298P, which contains two H bridges. The H bridge connections are complex, but they allow the shield to halt a motor, deliver PWM signals, and reverse the motor's direction. In addition, the shield can deliver more electrical power than the Mega can.

The capabilities of the Arduino programming environment can be extended with libraries. The first library discussed in this chapter is the Stepper library, which makes it possible to control stepper motors. This library has four functions, and two of them are constructors that return a `Stepper` object. After this object is created, its member functions—`setSpeed` and `step`—can be called.

The functions of the Servo library make it possible to control servomotors. To accomplish this in code, a sketch needs to declare a `Servo` object and associate it with a pin capable of delivering PWM signals. When you're using this library, it's important to know the minimum and maximum pulse widths needed to set the angle of the servomotor's shaft.

MOTOR CONTROL WITH THE RASPBERRY PI

The Raspberry Pi combines the processing resources of a personal computer with the small size of an Arduino board. In fact, the Raspberry Pi (shortened to *RPi*) is even smaller than the Arduino Mega discussed in the preceding chapter—and despite the RPi's superior capabilities, the two boards have approximately the same cost.

To be precise, the RPi is a *single-board computer*, or *SBC*. Instead of relying on a microcontroller, its data crunching is performed by a full processor: the Broadcom BCM2835. In addition, the RPi has enough memory to store a complete operating system.

In my opinion, the ability to support an operating system is the RPi's greatest strength. With an OS running on the device, I don't have to worry about low-level memory access—the system takes care of it for me. Also, if the operating system is based on Linux, as most RPi OSes are, I don't have to learn a new programming language. I can use code that I've written for other Linux systems.

Despite its many strengths, the RPi has three disadvantages that every maker should keep in mind:

- **Power**—The RPi B+ consumes 3 W, which is over five times the power required by most Arduino boards.

- **No analog input**—The RPi can deliver PWM signals, but it has no analog-to-digital converters capable of reading analog inputs.

- **Proprietary design**—The design files for the RPi are not freely available. Therefore, if you'd like to design a product similar to the RPi, you'll have to start from scratch.

None of these drawbacks affects the ability of the RPi to control motors. This chapter explains how to program the RPi and how to use the RaspiRobot board to control servomotors, brushed motors, and stepper motors. But first, I'd like to provide an overview of the RPi's circuit board and processor.

10.1 The Raspberry Pi

Despite occupying the same area as a credit card, the RPi has enough computational power to run an operating system, display video to a monitor, and communicate over Ethernet—all at the same time. These capabilities are made possible by the densely packed circuit board and the BCM2835. This section discusses both topics.

10.1.1 The Raspberry Pi Circuit Board

The first Raspberry Pi board was released in early 2012, and was called the Model A. The Model B+ provides more connectors and general-purpose input/output (GPIO) pins. This chapter focuses exclusively on the Model B+, and Figure 10.1 shows what it looks like.

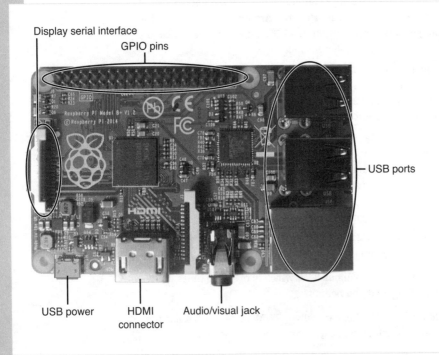

Figure 10.1
The Raspberry Pi
Model B+

Display serial interface

GPIO pins

USB ports

USB power   HDMI
connector   Audio/visual jack

Table 10.1 lists the board's basic characteristics.

Table 10.1 Technical Specifications of the Raspberry Pi Model B+

Parameter Name	Parameter Value
Dimensions	3.37 × 2.205 inches (85.6 × 56 mm)
Operating voltage	5 V
SDRAM	512 MB
Nonvolatile memory	MicroSD card
General-purpose input/output (GPIO) pins	40

In addition to its GPIO pins, the RPi has five USB connectors: one to provide power to the board and four to communicate with external devices. These USB ports make it possible to connect a keyboard and mouse, and the HDMI connector makes it possible to connect a monitor.

Because of these connections, the RPi can be accessed as a standalone personal computer. The RPi doesn't provide all the capabilities of a traditional PC, but it serves as an excellent way for users to run programs and interact with a Linux-based operating system. This interaction is made possible by the BCM2835, which is discussed next.

10.1.2 The BCM2835 System on a Chip (SoC)

The BCM2835 serves as the central brain of the Raspberry Pi. It isn't labeled in Figure 10.1 because it's stacked beneath the Samsung RAM chip in the center-left of the board.

Many sources refer to the BCM2835 as a processor, but this device contains two processing units: an ARM1176JZ-F general-purpose data processor and a dual-core VideoCore graphics processor. These processing units are properly called *cores*, and because the BCM2835 contains multiple cores, the proper term for the device is a *system on a chip*, or *SoC*.

To appreciate the RPi's computing power, it's good to be familiar with both cores of the BCM2835. This discussion provides a brief overview of the ARM1176 processing core and the VideoCore IV graphics processing core.

The ARM1176 Processing Core

When most people think about companies that make processors, they think of corporations such as Intel and AMD. These companies produce physical chips, such as the Core i7 and Athlon.

ARM Holdings plc also makes processors, but it doesn't manufacture physical devices. Instead, it sells processor designs to other companies, which incorporate them in their chips. The BCM2835 is a perfect example. Broadcom manufactures and sells the BCM2835 device, but the device's processor design, or core, was designed by ARM.

ARM cores are divided into families, and the ARM11 family was released in 2002. The processors in this family can operate on 32 bits at a time, and their execution speed ranges from 750 MHz up to 1 GHz. This family contains the ARM1176, which is the core that processes data in the Raspberry Pi.

One of the main benefits of using an ARM11 processing core is the availability of SIMD (single-instruction, multiple data) processing. Each ARM11 processor has a vector floating-point (VFP) coprocessor capable of performing arithmetic operations on multiple floating-point values at the same time. This coprocessor is vitally important for computational tasks involving audio and video.

The VideoCore IV Graphics Processing Core

When I started working with the Raspberry Pi, the feature that amazed me the most was its graphics. The user interface to the Raspbian operating system isn't just functional—its desktop has a level of polish that I never would have believed possible for a board as small (and as cheap!) as the Raspberry Pi.

This incredible capability is made possible by the dual-core VideoCore IV graphics processing core. Just as ARM Holdings plc designs cores that process data, VideoCore designs cores that process graphics. The computational characteristics of the VideoCore IV include the following:

- Displays graphics at 720p standard resolution

- Renders 25 million triangles per second

- Supports high-speed anti-aliasing for 2D rendering

- Renders graphics with 16-bit high dynamic range (HDR)

- Supports the full OpenGL-ES 1.1 and 2.0 standards

In my opinion, the last point is particularly impressive. I've spent years developing applications with the Open Graphics Library (OpenGL), but the target was always a high-performance graphics card requiring hundreds of watts. The fact that a low-power system such as the RPi supports full OpenGL-ES rendering leaves me astounded.

10.2 Programming the Raspberry Pi

Developing software for the RPi is similar to developing software for a regular computer, with one significant exception: The operating system must be downloaded, installed, and inserted. By *inserted*, I mean that the RPi accesses its operating system through a MicroSD card that must be inserted into the receptacle on the rear side of the board. This card must be purchased separately from the RPi, and you can find a list of supported cards at http://elinux.org/RPi_SD_cards.

Many operating systems have been ported to run on the RPi, including Fedora Linux (called Pidora), Arch Linux, and Debian (called Raspbian). The Raspberry Pi Foundation recommends Raspbian, which is optimized to run on the RPi. Raspbian provides nearly all of the features you'd expect from a Debian distribution and all the utilities you need to be productive.

This section starts by providing an overview of the Raspbian environment. The next part explains how to write and execute Python scripts, and how to access the board's general-purpose input/output (GPIO) pins in Python.

The last part of this section explains how to enable pulse width modulation (PWM) on the RPi. Once you understand how to generate PWM signals, you're well on your way to controlling motors.

10.2.1 The Raspbian Operating System

After the operating system is installed, the RPi can be connected to a monitor through the board's HDMI port. Figure 10.2 shows what Raspbian's initial desktop looks like.

Figure 10.2
The Raspbian desktop

Just as in Windows or Mac OS, applications can be executed by double-clicking icons in the Raspbian desktop. The following list presents a subset of the available applications:

- **Epiphany**—A scaled-down web browser

- **IDLE**—Development environment for Python coding (all versions less than 3.0)

- **LXTerminal**—A terminal for entering commands

- **IDLE3**—Development environment for Python 3.*x* coding

- **WiFi Config**—Configuration tool for Wi-Fi communication

- **Shutdown**—Shuts down the Raspberry Pi

By launching the terminal application, you can use all the utilities common to Linux distributions, such as ls, cd, cat, and rm. For this chapter, the only application you need to be familiar with is IDLE. The following discussion explains how it can be used to edit, compile, and run applications.

10.2.2 Python and IDLE

The Raspberry Pi development tools support a number of languages, including Python, C, C++, Java, and Ruby. The Raspberry Pi foundation recommends Python, which is also the most supported and widely used language. A full discussion of the Python language is beyond the scope of this book, but for beginners, I recommend the online book *Python for You and Me* at http://pymbook. readthedocs.org.

The RPi desktop provides two graphical utilities for editing Python: IDLE for Python 2.*x* and IDLE3 for Python 3.*x*. IDLE stands for Integrated DeveLopment Environment, and this chapter focuses on Python 2.*x*. Figure 10.3 shows what the IDLE dialog looks like.

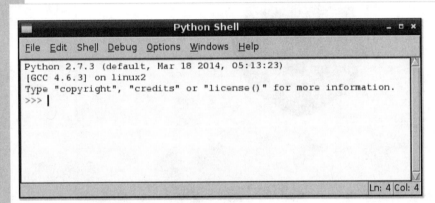

Figure 10.3
The Python Integrated DeveLopment Environment (IDLE)

This dialog provides access to a shell that allows you to enter Python commands. If you type `2+2` and press Enter, the shell will respond with 4. If you type `print Hello world`, the shell will respond with `Hello world`.

To take full advantage of the Python language, you can store statements in files called *scripts*. To create a script in IDLE, go to File > New File in the main menu or press Ctrl+N. This opens a second window that allows you to edit a script. Figure 10.4 shows what this window looks like.

In addition to syntax coloring, IDLE's editor window provides many standard capabilities for working with script code:

- To check the script for errors, go to Run > Check Module or press Alt+X.

- To save the script's content, go to File > Save or press Ctrl+S.

- To run the script's commands, go to Run > Run Module or press F5.

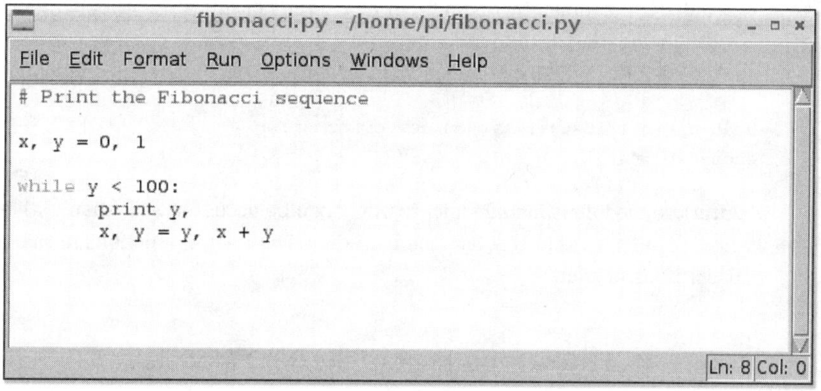

Figure 10.4
The IDLE editor window

```
fibonacci.py - /home/pi/fibonacci.py          _ □ ×

File  Edit  Format  Run  Options  Windows  Help

# Print the Fibonacci sequence

x, y = 0, 1

while y < 100:
        print y,
        x, y = y, x + y



                                                    Ln: 8 Col: 0
```

When a Python script is executed from the editor, the first window displays the output. It also displays any errors that arise during the script's execution.

Python makes it possible to access many of the RPi's resources, including the network connection, USB connections, and the graphical processor, but motor control requires access to the RPi's general-purpose input/output (GPIO) pins. I'll discuss this next.

10.2.3 Interfacing GPIO

At the top of the circuit board, the 40-pin header makes it possible to connect the RPi to external circuits. Twenty-six of the header's pins are accessible as GPIO pins. These pins are numbered from 2 to 27, and Figure 10.5 shows where they're located.

Figure 10.5
GPIO pins of the Raspberry Pi B+

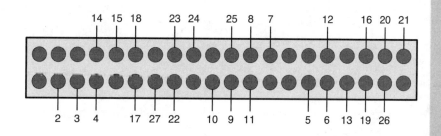

Two software modules make it possible to access these pins in Python:

- **RPI.GPIO**—This is available from http://sourceforge.net/projects/raspberry-gpio-python and is released under the MIT license.

- **RPIO**—This is available from http://pythonhosted.org/RPIO and is released under the GPL3 license.

Both modules make it possible to configure GPIO pins and read/write digital values. RPI.GPIO is installed in Raspbian by default, but at the time of this writing, only RPIO enables access to pulse width modulation at the hardware level. Because PWM is central to motor control, this discussion relies on RPIO, which can be installed by entering the following commands in a terminal:

```
sudo apt-get install python-dev python-pip
sudo pip install -U RPIO
```

After the module is installed, its features can be accessed in Python with a statement such as `import RPIO`. Table 10.2 lists many of the GPIO-related functions in this module and provides a description of each.

Table 10.2 GPIO Functions of the RPIO Module

Function	Description
`setmode(int num_mode)`	Identifies the method of pin numbering
`setup(int pin, int mode)`	Configures a pin as an input or output
`setup(int pin, int mode, int res_mode)`	Configures a pin as an input or output, and connects a pull-up or pull-down resistor
`output(int pin, int level)`	Sets a pin's logic level to `RPIO.HIGH` or `RPIO.LOW`
`int input(int pin)`	Reads the logic level at the given pin
`cleanup()`	Sets pins to default state

`add_interrupt_callback`
`(int pin, callback_func,`
`edge='both',`
`pull_up_down=RPIO.PUD_OFF,`
`threaded_callback=False`
`debounce_timeout_ms=None)`
	Associates a callback function with the given pin and an event that meets the specified criteria

`wait_for_interrupts`
`(threaded=False,`
`poll_timeout=1)`
	Halts processing until an interrupt occurs

`del_interrupt_callback` `(int pin)`	Used to remove callbacks associated with the pin

The first function in this table is particularly important. The RPi supports two numbering schemes for its GPIO pins:

- `RPIO.BOARD`—The numbers printed on the board (1 in the upper-left, 40 in the lower-right).

- `RPIO.BCM`—The numbers are set by how the GPIO pins are connected to the BCM2835.

To specify how the pin numbers should be interpreted, the `setMode` function accepts either value. The `RPIO.BCM` mode is the numbering scheme illustrated in Figure 10.5, and it makes it easier to interface external devices. For this reason, all the code in this chapter starts with the following line:

```
RPIO.setmode(RPIO.BCM)
```

The rest of the functions in Table 10.2 can be divided into two groups: those that involve input and output, and those related to events. The following discussion presents the functions in both groups.

Input and Output Pins

After the pin numbering method is set, the next step involves specifying which pins are input pins and which are output pins. This is accomplished with the `setup` function, which accepts a pin number and either `RPIO.IN` (input) or `RPIO.OUT` (output). As an example, the following code configures GPIO Pin 24 as an output pin:

```
RPIO.setup(24, RPIO.OUT)
```

If a pin is configured for output, its logic level can be set with the `output` function. This accepts the pin number and the logic level. If the logic level is set to `RPIO.HIGH` or 1, the pin's voltage will be set to 3.3 V. If the logic level is set to `RPIO.LOW` or 0, the pin's voltage will be set to 0 V.

If a pin is configured for input, its logic level can be read with the `input` function, whose only argument is the pin number. From my tests, this returns 1 if the pin's voltage is greater than 1.6 V and returns 0 if the voltage is less than 0.6–0.7 V. If the voltage is between 0.7 V and 1.6 V, `input`'s return value can't be determined.

The code in Listing 10.1 shows how `setup`, `input`, and `output` are used in practice. This script reads from GPIO Pin 17 and sets the value of GPIO Pin 24.

Listing 10.1   Ch10/check_input.py—Checking a Pin's Logic Level

```
"""
This code repeatedly checks the logic level of in_pin.
If the level is low, out_pin is set high and the reading continues.
If the level is high, the script completes.
"""

import RPIO

# Set input pins
in_pin = 17;
out_pin = 24;

# Specify use of BCM pin numbering
RPIO.setmode(RPIO.BCM)
```

```
# Configure pin directions
RPIO.setup(in_pin, RPIO.IN)
RPIO.setup(out_pin, RPIO.OUT)

# Wait for in_pin to reach low voltage
while(RPIO.input(in_pin) == RPIO.LOW):
  RPIO.output(out_pin, RPIO.HIGH)

# Return pins to default state
RPIO.cleanup()
```

In this script, each iteration of the `while` loop checks the state of the input pin. If the pin's level is low, the voltage of the output pin is set high. The loop continues until the input pin's voltage is set high.

By default, the logic level of an input pin is floating, which means it may randomly take high or low values. The initial logic level can be controlled in code by adding an optional third argument to `setup`. This argument sets the `pull_up_down` variable to one of three values:

- `RPIO.PUD_UP`—Connects a pull-up resistor to the pin.

- `RPIO.PUD_DOWN`—Connects a pull-up resistor to the pin.

- `RPIO.PUD_OFF`—Doesn't connect the pin to a pull-up or pull-down resistor. This is the default state.

As an example, the following code configures Pin 17 to serve as an input pin connected to power through a pull-up resistor:

```
RPIO.setup(17, RPIO.IN, pull_up_down=RPIO.PID_UP)
```

In Listing 10.1, the final function executed in the script is `cleanup`. This returns each GPIO pin to its default configuration and logic level.

Handling Interrupts

The last three functions in Table 10.2 relate to interrupts. An *interrupt* is a state change that may cause the processor to stop what it's doing, execute a routine to handle the interrupt, and then return to its processing. RPIO makes it possible to handle two types of interrupts: networking interrupts and GPIO interrupts. This discussion is concerned with GPIO interrupts.

The `add_interrupt_callback` function tells RPIO that a pin should be monitored for a specific type of event. This function accepts six arguments and the last four are optional:

- The GPIO pin to be monitored.

- The name of the function to handle the interrupt.

- The type of logic level change:

  - `rising`—The pin's logic changes from low to high (rising edge).

  - `falling`—The pin's logic changes from high to low (falling edge).

  - `both`—Either rising or falling.

- The input pin's connection to the power/ground:

  - `RPIO.PUD_UP`—Connects a pull-up resistor to the pin.

  - `RPIO.PUD_DOWN`—Connects a pull-down resistor to the pin.

  - `RPIO.PUD_OFF`—Leaves the pin's logic level floating.

- The handling function's execution in a thread:

  - `true`—RPIO calls the interrupt-handling function in a separate thread.

  - `false`—The current program halts to execute the interrupt handling function.

- The minimum number of seconds allowed between interrupts.

Only the first two arguments are required, and the second argument identifies which function should be called to handle the interrupt. This function is referred to as a *callback function* or just a *callback*, and RPIO passes two arguments to it: the pin number and an integer that identifies whether the event involved a rising edge (1) or a falling edge (0).

Interrupts and callbacks can be hard to understand, so Listing 10.2 shows how they're configured and used in code. This script monitors Pin 17 and calls a different callback depending on the event.

Listing 10.2　Ch10/interrupt.py—Responding to Logic Level Changes

```
"""
This code sets up interrupt handling for Pin 17.
A change in the logic level executes a callback
that prints a message.
"""

import RPIO

def edge_detector(pin_num, rising_edge):
  if rising_edge:
    print("Rising edge detected on Pin %s" % pin_num)
  else:
    print("Falling edge detected on Pin %s" % pin_num)

# Define input pin
in_pin = 17

# Specify use of BCM pin numbering
RPIO.setmode(RPIO.BCM)
```

```
# Configure pin direction
RPIO.setup(in_pin, RPIO.IN)

# Configure interrupt handling for rising and falling edges
RPIO.add_interrupt_callback(in_pin, edge_detector, edge='both')
RPIO.wait_for_interrupts()

# Return pin to default state
RPIO.del_interrupt_callback(in_pin)
RPIO.cleanup()
```

After setting the callback with `add_interrupt_callback`, the script calls `wait_for_interrupts`. If called without arguments, this function halts processing until an interrupt occurs. However, if the first argument sets `threaded` to TRUE, the waiting will be performed in a background thread. The following code shows how this is used:

```
RPIO.wait_for_interrupts(threaded=TRUE)
```

The last function in Table 10.2 is `del_interrupt_callback`. This accepts a pin number and removes all callbacks assigned to that pin.

10.3 Controlling a Servomotor

As discussed in Chapter 5, "Servomotors," a hobbyist servomotor is controlled with three pins: power (usually 5–6 V), ground, and control. The control pin sets the angle of the servomotor's shaft by providing a train of pulses formatted according to pulse width modulation (PWM).

Chapter 2, "Preliminary Concepts," explained the basics of PWM. This section explains how to generate a PWM signal on the RPi and how to use this to control a servomotor.

10.3.1 Configuring PWM

As discussed in Chapter 2, pulse width modulation (PWM) controls motors by generating pulses of varying width but spaced at a constant interval (the period). RPIO includes a PWM module whose functions generate pulses by accessing the processor's underlying direct memory access (DMA) capability.

This low-level DMA access makes it possible to generate PWM signals without interrupting regular processing. You don't have to understand how DMA works on the processor, but there are two vital facts to keep in mind:

- RPIO provides access to 15 DMA channels, numbered from 0 to 14.

- A DMA channel can be associated with one or more GPIO pins. After the association is made, it will be able to deliver pulses to those pins with high precision and resolution.

Table 10.3 lists many, but not all, of the PWM functions provided by RPIO.PWM.

Table 10.3 Functions of the RPIO.PWM Module

Function	Description
`setup(pulse_incr_us=10, delay_hw=0)`	Initializes the DMA channels for use
`init_channel(int dma_channel, subcycle_time_us=20000)`	Configures the cycle with a specific period (20 ms by default)
`add_channel_pulse (int dma_channel, int pin, int start, int width)`	Generates pulse of the given width for the pin
`clear_channel (int dma_channel)`	Clears all pulses from the channel
`clear_channel_gpio (int dma_channel, int pin)`	Clears the pulse for the specified pin from the channel
`cleanup()`	Halts PWM and DMA

The first function, `setup`, must be called before any of the other functions in the table. Both of its parameters are optional, and the first identifies the desired pulse width resolution. I recommend setting this value to 1. The second parameter can be set to `PWM.DELAY_VIA_PWM` or `PWM.DELAY_VIA_PCM`.

After `setup` initializes the DMA operation, `init_channel` configures a specific channel for use. The first parameter identifies the channel number, and may be set to any number between 0 and 14. The second sets the desired time between the pulses' rising edges. The default time is set to 20 ms, which is the pulse interval expected by every servo I've encountered.

The most important function in the table is `add_channel_pulse`, which generates pulses for a particular pin. The first two parameters identify the DMA channel and the GPIO pin. The third and fourth parameters characterize the shape of the pulses. The `start` parameter identifies the offset of the pulse relative to the start of the period, and the `width` parameter identifies the pulse's width. Figure 10.6 gives an idea what this look like.

Figure 10.6
Pulses generated by
RPIO.PWM

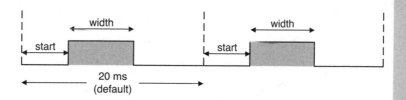

The `start` parameter may not seem useful at first because the same time offset is applied to every pulse, but it's helpful when you want to generate multiple pulses within a single period. For example, the following code generates two pulses for DMA Channel 5 and GPIO Pin 18. Both pulses have a width of 2 ms and the second pulse is delayed by 8 ms.

```
PWM.add_channel_pulse(0, 18, 0, 1000)
PWM.add_channel_pulse(0, 18, 8000, 1000)
```

The last three functions in Table 10.3 become important when PWM signaling is no longer needed. The `clear_channel` function clears all pulses from the given DMA channel, and `clear_channel_gpio` clears the DMA pulses for a specific GPIO pin. The `cleanup` method halts all PWM and DMA operation.

The code in Listing 10.3 demonstrates how to generate PWM signals on the Raspberry Pi. It initializes DMA Channel 0 and delivers pulses to GPIO Pin 18. Each pulse has a width of 1000 ms.

Listing 10.3   Ch10/pwm.py—Generating a PWM Signal

```
"""
This code generates a pulse-width modulation (PWM)
for Pin 18 whose pulses have a width of 1ms.
"""

import RPIO.PWM as PWM
import time

# Define PWM pin
pwm_pin = 18

# Initialize DMA and set pulse width resolution
PWM.setup(1)

# Initialize DMA channel 0
PWM.init_channel(0)

# Set pulse width to 1000us = 1ms
PWM.add_channel_pulse(0, pwm_pin, 0, 1000)

time.sleep(10)

# Clear DMA channel and return pins to default settings
PWM.clear_channel(0)
PWM.cleanup()
```

As the program executes on the RPi, RPIO prints the following status messages to the console:

```
Using hardware: PWM
PW increments: 1us
Initializing channel 0...
add_channel_pulse: channel=0, start=0, width=1000
init_gpio 18
```

Generating the PWM signal isn't enough—the program must be delayed so that it delivers multiple pulses to the output pin. As demonstrated in the code, this can be accomplished with the `sleep` function of the time module. This accepts a delay value in seconds, so `sleep(10)` delays the program's execution for 10 seconds.

10.3.2 Controlling a Servo

The PWM module provides access to a class called `Servo`, which provides two methods specifically suited for controlling servomotors:

- `set_servo(int pin, int width)`—Delivers a PWM signal through the specified pin whose pulses have the specified width (in microseconds)

- `stop_servo(int pin)`—Halts the PWM signal for the given pin

The main advantage of using these methods over the regular PWM methods is that there's no need to initialize DMA or set the appropriate DMA channel. That is, there's no reason to call `PWM.setup` or `PWM.init_channel`.

To see how these functions can be used to control a real servomotor, consider the FS5106B from Fitec. Like many servomotors, it expects control pulses to have their rising edges spaced 20 ms apart. The shaft rotates to the minimum angle when the pulse width equals 0.7 ms and rotates to its maximum angle when the pulse width equals 2.3 ms. A pulse width of 1.5 ms sets the shaft to its neutral position.

The code in Listing 10.4 shows how `set_servo` and `stop_servo` can be used to control the Fitec FS5106B. This program rotates the servo's shaft from the minimum angle to the maximum angle, and then returns to the minimum angle.

Listing 10.4   Ch10/servo.py—Control a Servomotor

```
"""
This code controls a servomotor, rotating from
the minimum to maximum angle and back.
"""

import RPIO.PWM as PWM
import time

# Define control pin and pulse widths
servo_pin = 18
min_width = 700
max_width = 2300

# Create servo object
servo = PWM.Servo()

# Set the angle to the minimum angle and wait
servo.set_servo(servo_pin, min_width)
time.sleep(1)

# Rotate shaft to maximum angle
for angle in xrange(min_width, max_width, 100):
  servo.set_servo(servo_pin, angle)
  time.sleep(0.25)
```

```
# Rotate shaft to minimum angle
for angle in xrange(max_width, min_width, -100):
  servo.set_servo(servo_pin, angle)
  time.sleep(0.5)

# Stop delivering PWM to servo
servo.stop_servo(servo_pin)
```

Every time the `set_servo` method is called, `time.sleep` is called afterward to delay the program. This is necessary to give the motor enough time to reach the desired angle.

10.4 The RaspiRobot Board

Due to the RPi's worldwide popularity, many makers have designed expansion boards that extend its capabilities. The PiFace board provides switches, relays, and LEDs. The PiRack has headers that make it possible to connect other circuit boards to the RPi. If you want to see what types of expansion boards are available, a good place to look is https://www.modmypi.com/raspberry-pi-expansion-boards.

The RaspiRobot board is an expansion board whose components enable the RPi to control motors. Figure 10.7 shows what it looks like.

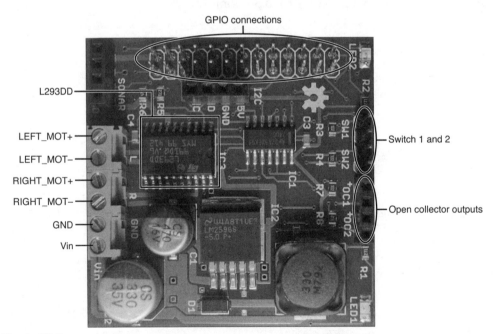

Figure 10.7
The RaspiRobot expansion board

The six screw terminals on the left are particularly important. The upper four connections (LEFT_ MOT+, LEFT_MOT-, RIGHT_MOT+, and RIGHT_MOT-) can deliver power to two brushed DC motors or one stepper motor. The lower connections (GND and Vin) provide power to the RaspiRobot board. Vin must be connected to 7–12 volts of DC power.

To take advantage of the RaspiRobot's capabilities, you need to know how its peripherals are connected to the Raspberry Pi's GPIO pins. Table 10.4 lists the RaspiRobot's signal names and their corresponding GPIO pins.

Table 10.4 RaspiRobot Signals and GPIO Pins

RaspiRobot Signal	RPi GPIO Pin	Description
LEFT_GO_PIN	17	PWM signal for left motor
LEFT_DIR_PIN	4	Controls the direction of left motor
RIGHT_GO_PIN	10	PWM signal for left motor
RIGHT_DIR_PIN	25	Controls the direction of left motor
SW1_PIN	11	Connected to Switch 1
SW2_PIN	9	Connected to Switch 2
LED1_PIN	7	Connected to LED 1
LED2_PIN	8	Connected to LED 2
OC1_PIN	22	Open collector output 1
OC2_PIN	27	Open collector output 2
OC2_PIN_R1	21	Open collector resistor output 1
OC2_PIN_R2	27	Open collector resistor output 2
TRIGGER_PIN	18	Transmits trigger pulse for sonar
ECHO_PIN	23	Receives sonar echo

The first four signals provide motor control. They're connected to the board's central integrated circuit, the L293DD Quadruple Half-H Driver. This device produces the LEFT_MOT+, LEFT_MOT-, RIGHT_MOT+, and RIGHT_MOT- signals that drive the motors. This section explains how this device works and then discusses the process of coding applications that control brushed DC motors and stepper motors.

10.4.1 The L293DD Quadruple Half-H Driver

Chapter 2 explained how an H bridge makes it possible to deliver current to a motor in the forward and reverse directions. H bridges can be implemented with discrete transistors, but many circuits rely on integrated circuits. The L293 is one of the most popular ICs for this purpose, and many companies sell their own variants.

The RaspiRobot board contains the L293DD device from ST Microelectronics. This 20-pin surface-mount chip contains four half-H bridges, which can be connected to form two full H bridges. The device receives four inputs (LEFT_GO_PIN, LEFT_DIR_PIN, RIGHT_GO_PIN, and RIGHT_DIR_PIN) and delivers four outputs (LEFT_MOT+, LEFT_MOT-, RIGHT_MOT+, and RIGHT_MOT-) that provide power to the RaspiRobot's motor connections.

Figure 10.8 shows how one of the L293DD's H bridges is connected. In this schematic, the RaspiRobot directs LEFT_GO_PIN and LEFT_DIR_PIN to an H bridge that produces LEFT_MOT+ and LEFT_MOT-.

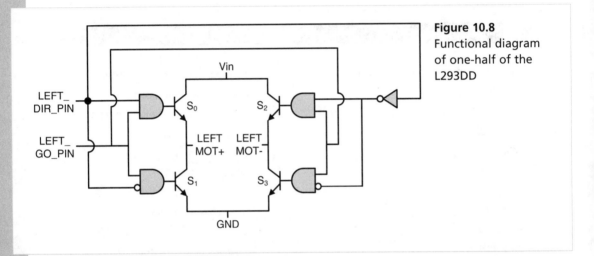

Figure 10.8
Functional diagram of one-half of the L293DD

If LEFT_GO_PIN is low, none of the switches (S_0 through S_3) can receive any input voltage. For this reason, LEFT_GO_PIN is driven by a PWM signal from the controller. The higher the PWM duty cycle, the longer the switches are closed.

If LEFT_GO_PIN is high, LEFT_DIR_PIN determines the values of LEFT_MOT+ and LEFT_MOT-. If LEFT_DIR_PIN is high, S_0 and S_3 are closed, setting LEFT_MOT+ equal to Vin and LEFT_MOT- equal to GND. If LEFT_DIR_PIN is low, S_1 and S_2 are closed, setting LEFT_MOT+ equal to GND and LEFT_MOT- equal to Vin.

10.4.2 RaspiRobot Python Code

The RaspiRobot's features can be accessed through the GPIO and PWM functions discussed earlier, but there's an easier way. Simon Monk, the designer of the RaspiRobot, provides a Python module called rrb2.py that simplifies developing code for the RPi and RaspiRobot. This code can be freely downloaded from https://github.com/simonmonk/raspirobotboard2.

This source file defines a class called RRB2, which contains a wide range of methods that access the RaspiRobot's features. Some methods access switches and others activate the board's LEDs. Six of the class's methods make it possible to control the operation of two brushed motors–one on the left and one on the right. These methods are listed in Table 10.5.

Table 10.5   Motor Control Methods of the RRB2 Class

Function	Description
`forward(seconds=0, speed=0.5)`	Drives the motor forward at the given speed for the specified amount of time
`reverse(seconds=0, speed=0.5)`	Drives the motor in reverse at the given speed for the specified amount of time
`left(seconds=0, speed=0.5)`	Drives the motor forward at the given speed for the specified amount of time
`right(seconds=0, speed=0.5)`	Drives the motor in reverse at the given speed for the specified amount of time
`stop()`	Halts the motor
`set_motors (float left_pwm, int left_dir, float right_pwm, int right_dir)`	Specifies the pins to be used for motor control

The first five methods accept the same two arguments: the time in seconds and the speed, which corresponds to the PWM duty cycle. The `speed` parameter must be given a value between 0.0 (no power to motors) and 1.0 (full power to motors).

The last method in the table, `set_motors`, provides direct access to the two motors. The first pair of parameters sets the duty cycle and direction of the left motor, and the second pair sets the duty cycle and direction of the right motor. A direction value of 0 rotates in the forward direction and a value of 1 rotates in the reverse direction.

The `RRB2` class also contains member variables that correspond to signals of the RaspiRobot board. These have the same names as the signals listed in the left column of Table 10.4. For example, rather than remember that Pin 25 controls the direction of the right motor, you can use `RRB2.RIGHT_DIR_PIN`, which equals 25.

10.4.3 Controlling Brushed DC Motors

The methods of the `RRB2` class make it easy to control two brushed DC motors, so long as the left motor is connected to LEFT_MOT+ and LEFT_MOT- and the right motor is connected to RIGHT_MOT+ and RIGHT_MOT-. Listing 10.5 shows how two such motors can be controlled through the RaspiRobot. This code drives the motors forward for 5 seconds, backwards for 4 seconds, right for 3 seconds, and left for 2 seconds.

Listing 10.5   Ch10/brushed.py—Control Two Brushed DC Motors

```
"""
This program controls two brushless DC motors:
Forward for five seconds, backwards for four seconds,
right for three seconds, and left for two seconds.
"""
```

```
import rrb2

# Create RRB2 object
robot = rrb2.RRB2()

# Rotate forward for five seconds
robot.forward(seconds=5, speed=1.0)

# Rotate backward for four seconds
robot.reverse(seconds=4, speed=0.8)

# Turn left for three seconds
robot.left(seconds=3, speed=0.6)

# Turn right for two seconds
robot.right(seconds=2, speed=0.4)

# Stop motor
robot.stop()
```

This code is straightforward to understand. It starts by creating an RRB2 object, and it's worth noting that the RRB2 constructor accepts a `revision` value that identifies the RaspiRobot version. By default, this is set to 2.

10.4.4 Controlling a Stepper Motor

The RaspiRobot code doesn't provide any functions specifically for stepper motors, but the `setMotors` method in Table 10.5 can serve the purpose. But before I get into the code, I'd like to review a few basic concepts related to steppers. Chapter 4, "Stepper Motors," provides a full discussion of this topic.

Basics of Stepper Control

Stepper motors come in two types: unipolar and bipolar. Unipolar steppers have four connections and are more efficient than bipolar steppers, but they require more complex control circuitry (two H bridges). A bipolar stepper can be connected as a unipolar motor by leaving its power wires unconnected.

The RaspiRobot board contains two H bridges and has only four terminals available for controlling motors. Therefore, this discussion assumes that the target stepper is a unipolar stepper or a bipolar stepper connected to run as a unipolar stepper.

Stepper motors have two phases, A/A' and B/B'. For a stepper, these windings must be driven so that the rotor is attracted to one phase and repelled by the other, and then repelled by the first phase and attracted to the second. Figure 10.9 shows how this works.

Figure 10.9
Operation
of a stepper
motor

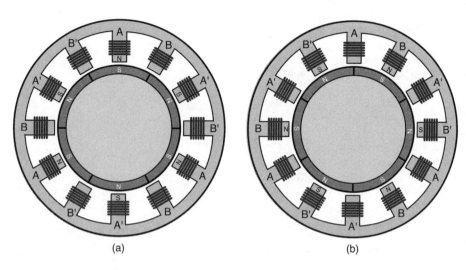

(a)   (b)

In Figure 10.9a, the A/A' windings are energized in such a way that A behaves as a north pole and A' behaves as a south pole. The B/B' windings are left unenergized. In Figure 10.9b, the situation is reversed: the A/A' windings are left unenergized and the B/B' windings are energized so that B behaves as a north pole and B' behaves as a south pole. As a result, the rotor steps through an angle of 30°.

Controlling a Stepper with the RRB2 Class

To see how the RaspiRobot can control a stepper, you need to think of the A/A' windings as the left motor and the B/B' windings as the right motor. Then, if setMotors is called with the right values in the right sequence, the stepper will rotate through its step angle. The code in Listing 10.6 shows how this can be done.

Listing 10.6   Ch10/stepper.py—Control a Stepper Motor

```
"""
This program controls a stepper motor by
energizing its phases in a given sequence.
"""

import rrb2
import time

# Create RRB2 object
robot = rrb2.RRB2()
```

```
# Set number of repetitions and step delay
num_reps = 10
step_delay = 0.4

# Repeat the energizing sequence num_reps times
for x in range(0, num_reps):

  robot.set_motors(1.0, 1, 0.0, 0)
  time.sleep(step_delay)

  robot.set_motors(0.0, 0, 1.0, 1)
  time.sleep(step_delay)

  robot.set_motors(1.0, 0, 0.0, 0)
  time.sleep(step_delay)

  robot.set_motors(0.0, 1, 1.0, 0)
  time.sleep(step_delay)

# Stop motor
robot.stop()
```

In this code, the PWM duty cycle is set to 1.0 when the winding is energized and 0.0 when it isn't. When the duty cycle is 1.0, the direction value (0 or 1) sets the polarity of the corresponding winding.

10.5 Summary

The Raspberry Pi packs a great deal of functionality into a tiny form factor. Thanks to the BCM2835, it can process graphics and general-purpose data at high speed. Its connectors and external devices make it possible to connect to the RPi as though it was a regular personal computer, complete with an Ethernet connection, HDMI connection, and multiple USB ports.

This chapter is primarily focused on the RPi's ability to control motors. In presenting this subject, I've had to make a number of decisions: Raspbian as the operating system, Python as the programming language, IDLE as the text editor, and RPIO as the GPIO-control module. Other choices are available in each case, but these options are easy to work with and very well supported.

The Model B+ version of the RPi provides 40 GPIO pins. The RPIO module makes it possible to specify whether they read input or write output. Programs can respond to changes to a pin's state by configuring interrupts and interrupt-handling routines. If the program assigns a callback function for an interrupt, that function will be called if the corresponding event takes place.

RPIO.PWM makes it straightforward to generate pulse width modulated (PWM) signals. To make life even simpler, the `Servo` class in RPIO.PWM provides two methods, `setServo` and `stopServo`, that make it easy to control servomotors through PWM.

Simon Monk's RaspiRobot board extends the RPi's capabilities by enabling it to control brushed DC motors and stepper motors. This control is made possible by the L293DD. The H bridges in this device receive input from the RPi's GPIO pins and deliver power to the motor(s) connected to the RaspiRobot board.

Instead of controlling the RaspiRobot's inputs directly, it's easier to call functions in the rrb2 module. This defines a class named RRB2 whose methods drive two brushed DC motors forward or in reverse. The `set_motors` method of this class provides control over the duty cycle and direction of each motor. For this reason, it's ideal for controlling the four inputs of a single stepper motor.

CONTROLLING MOTORS WITH THE BEAGLEBONE BLACK

The BeagleBone Black (BBB) is a single-board computer similar to the Raspberry Pi discussed in the previous chapter. The two computers have a lot in common: Both provide a wide range of capabilities on a board the size of a credit card. Both process data with an ARM processing core, both have 512 MB of RAM, and both have enough resources to run a full operating system. They're also approximately the same price ($35 for the Raspberry Pi, $55 for the BeagleBone Black).

Unlike the Raspberry Pi, the BBB contains a number of peripheral cores that serve roles similar to those of microcontrollers. These cores enable the BBB to convert analog signals to digital data and generate high-precision pulse width modulation (PWM) signals. When it comes to controlling motors, these are important capabilities.

The BBB's capabilities can be extended with an expansion board called the Dual Motor Controller Cape, or DMCC. This provides many of the same features as the add-on boards presented in Chapter 9, "Motor Control with the Arduino Mega," and Chapter 10, "Motor Control with the Raspberry Pi." In addition, it contains a processor that reads a motor's position and computes the feedback needed to set the motor's speed. The last part of this chapter explains how to use the DMCC to control DC motors.

However, before delving into motor control, it's important to be familiar with the capabilities of the BeagleBone Black. This chapter begins by discussing the board's features and the Debian operating system. I'll also present the fundamentals of BBB programming with Python and show how to access its general-purpose input/output (GPIO) pins.

11.1 The BeagleBone Black (BBB)

The BBB's circuit board provides a great deal of connectivity through its Ethernet, USB, HDMI, and MicroSD connectors. To access this connectivity in an application, you need to write code for the AM3359, which serves as the board's central device. This section discusses the board and the AM3359.

11.1.1 The BBB Circuit Board

In 2008, Texas Instruments released the first BeagleBoard to showcase the computational power of its processing devices. In 2011, the company released a second version of the BeagleBoard with a more powerful processor and many more peripheral connections.

In 2013, TI released a version of the BeagleBoard called the BeagleBone Black. This provides a higher clock speed than the previous version and twice as much memory at half the price. This chapter focuses on Revision C of the BBB, and Figure 11.1 shows what its front side looks like.

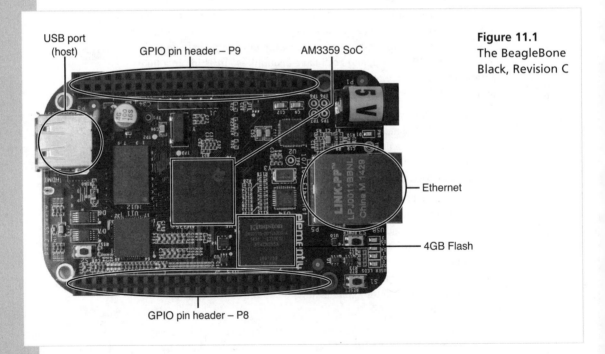

Figure 11.1
The BeagleBone
Black, Revision C

Table 11.1 lists the board's basic characteristics.

Table 11.1 Technical Specifications of the BeagleBone Black

Parameter	Value
Dimensions	3.402 × 2.098 inches (86.40 × 53.3 mm)
Operating voltage	5 V
RAM	512 MB DDR3
Nonvolatile memory	4 GB Flash memory and MicroSD card
General-purpose input/output (GPIO) pins	65

The BBB has two USB ports. The port on the front side makes it possible to connect as a host, which enables the BBB to access devices such as keyboards and mice. The port on the rear is used to connect as a device, which allows the BBB to be connected to a host such as a personal computer. The board can receive power through this rear connection or through the 5 V DC power jack.

With only one USB connection for devices, the BBB can't be immediately accessed as a standalone computer. However, if the BBB is connected to a USB hub, a keyboard and mouse can be connected through the hub.

The design files for the BeagleBone Black are freely available for download. This makes it possible to construct similar boards for your electronic projects. The schematics and design files can be obtained from the BBB wiki at http://elinux.org/Beagleboard:BeagleBoneBlack.

11.1.2 The AM3359 System on a Chip (SoC)

The central device of the BBB, the AM3359, is called a system on a chip, or SoC. This is because it has multiple processing cores on a single device. To be specific, the AM3359 contains an ARM-based processing core (Cortex-A8) and a dedicated core for processing graphics (SGX530). It also has a subsystem made up of two real-time processing cores that serve many of the same roles as microcontrollers.

The Sitara Cortex-A8 Processing Core

As discussed in Chapter 10, ARM Holdings plc sells processor designs to companies who integrate them in their chips. Texas Instruments purchased an ARM core for the BBB and rebrands it as the Sitara core. This core processes 32-bit data, runs at low power, and is capable of operating on multiple floating-point values at once using SIMD (single-instruction, multiple data) instructions.

To be specific, the AM3359's ARM core is a Cortex-A8 core. Its most important advantage is its dual-issue design. This enables the Cortex-A8 to process twice as many instructions in a given time as earlier ARM cores.

Another strength of the Cortex-A8 is its ability to execute NEON instructions. These instructions are specifically geared toward high-speed mathematics, and they operate on multiple values at once. Thanks to its NEON support, the BBB can crunch numbers at remarkably high speed.

The SGX530 3D Graphics Engine

The AM3359 contains a core specifically designed for graphics processing. This SGX530 core was designed by PowerVR, which is now a division of Imagination Technologies. This is the same graphics core used by the iPhone 4.

The processing characteristics of the SGX530 are given as follows:

- Displays graphics at 720p standard resolution

- Renders 14 million triangles per second

- Processes 200 million pixels per second

- Supports the full OpenGL-ES 1.1 and 2.0 standards

The SGX530 has many strengths, but it's not capable of decoding video. Therefore, the BBB relies on the Cortex-A8 to deliver output to the HDMI connector. For this reason, the BBB isn't suitable for graphics-intensive applications such as games.

The Programmable Real-Time Unit Subsystem and Industrial Communication Subsystem (PRU-ICSS)

Within the AM3359, the PRU-ICSS contains two real-time cores called programmable real-time units, or PRUs. These PRUs are 32-bit processors, but they can't execute all the instructions you'd expect from a full processor. Each has 8 KB of program memory and 8 KB of data memory.

The primary purpose of the PRU-ICSS is to handle basic input/output processing for the BBB, thereby freeing the ARM core to work on higher-level tasks. To serve this purpose, the PRU-ICSS contains its own Ethernet processing capability, a universal asynchronous receiver/transceiver (UART), and a dedicated interrupt controller for responding to external events.

11.2 Programming the BBB

Thanks to the ARM processor, the process of building applications for the BBB is similar to that for any ARM-based system. In fact, programming the BBB is easier because there are so many free resources online. One good place to look for information and support is http://beagleboard.org/Support/bone101.

But before you write code, you need to be familiar with the operating system. This section starts by providing an overview of the Debian OS, which comes preinstalled on the BBB.

Afterward, I'll explain how to write and execute Python scripts, and I'll show how the Adafruit_BBIO library makes it possible to access the board's general purpose input/output (GPIO) pins in Python. The last part of this section explains how to code applications that generate pulse width modulation (PWM) signals on the BBB.

11.2.1 The Debian Operating System

With its high-powered processor and generous amount of memory, the BeagleBone Black can run a full operating system. All of the OSes I've encountered are based on Linux: Many users run Debian and some run Ubuntu. Another popular operating system is Ångström, a Linux distribution developed specifically for embedded devices.

Some versions of the BBB have Ångström preinstalled, but current versions have Debian. From a developer's perspective, the differences between them are slight. One important point is that Ångström relies on opkg for package management and Debian relies on apt-get.

The MicroSD connector on the rear of the board makes it possible to install a new operating system, but this chapter focuses on the preinstalled Debian operating system. Figure 11.2 shows what the desktop looks like on my version of the BeagleBone Black.

Figure 11.2
The Debian
desktop

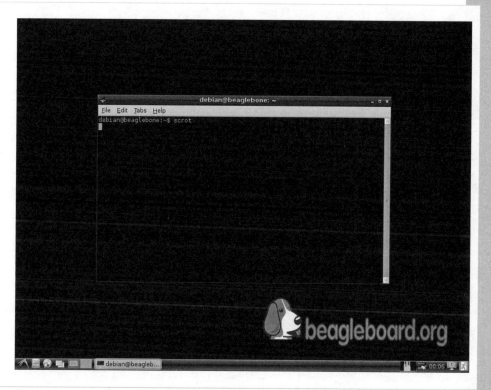

The default desktop doesn't have any icons, so users initially have to navigate using keystrokes or by clicking items in the menu at the bottom of the screen. I prefer to interact with the BBB through a secure shell (SSH), which allows me to enter and execute commands from my personal computer. When using SSH, there are two points to be aware of:

- The BBB's default IP address is 192.168.7.2.

- The BBB has a root account that doesn't require a password.

Combining this information, a user can launch an SSH session with the following command:

```
ssh root@192.168.7.2
```

In addition to entering commands via SSH, I prefer to transfer files using secure copy (SCP). The following command transfers data.txt from the current directory on my development system to the BBB's root directory:

```
scp data.txt root@192.168.7.2:/root
```

This command transfers /root/data.txt to the /home/matt folder on my development system:

```
scp root@192.168.7.2:/root/data.txt /home/matt
```

11.2.2 The Adafruit-BBIO Module

When I started working with the BBB, I wanted to program the AM3359 at a low level to obtain maximum performance and full configuration options. I learned about pin control multiplexing, device tree overlays, and all the registers the AM3359 uses to configure its operation. I even coded assembly language routines to run on the PRU-ICSS. In the end, I decided that low-level access to the BBB isn't worth the effort.

Thankfully, our friends at Adafruit developed a Python package that simplifies the development process. It's called the Adafruit_BBIO library and it's almost exactly similar to the RPIO module discussed in the preceding chapter. This might be preinstalled on your board, but in case it isn't, the following commands will install everything you need:

```
sudo apt-get update
sudo apt-get install build-essential python-dev python-setuptools
sudo apt-get install python-pip python-smbus -y
sudo pip install Adafruit_BBIO
```

After you're finished, you'll have all the tools needed to start coding Python scripts that use Adafruit_BBIO. You can execute a script on the command line by entering **python** followed the script's name. For example, the following command executes test.py:

```
python test.py
```

There's one last point I'd like to mention. In addition to C, C++, and Python, the BBB supports a fascinating language called BoneScript. This is a variation of JavaScript that makes use of the node.js framework. It was specifically developed to run on the BBB, and there are many tutorials and free resources on the Web. I chose Python for this chapter because the DMCC, discussed later in this chapter, can be accessed through Python but not BoneScript.

11.2.3 Accessing GPIO Pins

To connect to external circuits, the BBB has 92 pins that can be accessed through two 46-pin headers, P8 and P9. Each pin has a name that identifies its location in the header, such as P8.5. Some of these pins are configured for GPIO, and these pins have names that identify their position in the GPIO sequence. Figure 11.3 presents labels for the 24 leftmost pins in the P8 header.

Figure 11.3
GPIO pins of the
BeagleBone Black
(Header P8)

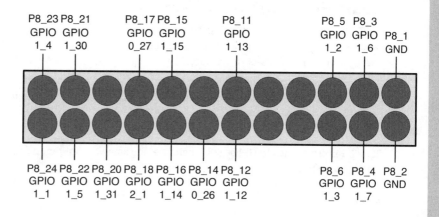

If you're willing to write C code and device tree overlays, you can configure GPIO pins in many different ways. However, if you only want to read or set the pins' voltage, the Adafruit_BBIO library is ideal. Table 11.2 lists seven of its functions.

Table 11.2 GPIO Functions of the Adafruit_BBIO Module

Function	Description
setup(string pin, int mode)	Configures a pin as an input or output
output(string pin, int level)	Sets the logic level to HIGH or LOW
int input(string pin)	Reads the logic level at the given pin
cleanup()	Sets the pins to the default state
wait_for_edge(string pin, int event)	Halts processing until the specified event occurs on the specified pin
add_event_detect(string pin, int event)	Watches for the specified event on the specified pin
event_detected(string pin)	Returns whether the watched event has occurred on the pin

These functions can be divided into two categories. The first contains functions related to basic pin configuration. The second contains functions that relate to events and event handling.

note

The code in this chapter imports the Adafruit_BBIO.GPIO module as GPIO and the Adafruit_BBIO.PWM module as PWM. Therefore, when discussing constants and functions in these modules, this chapter uses GPIO.*name* and PWM.*name*.

Basic Pin Configuration

GPIO pins can be configured for input (reading the pin's voltage level) or output (setting the pin's voltage level). This configuration is performed by the `setup` function, which accepts a pin string followed by either `GPIO.IN` or `GPIO.OUT`.

A pin string can be given as the header location (`"P8_12"`) or as the GPIO number (`"GPIO1_12"`). For example, the following code configures Pin 14 in Header 8 as an output pin:

```
GPIO.setup("P8_14", GPIO.OUT)
```

Pin P8_14 corresponds to GPIO0_26. Therefore, the same configuration can be accomplished with the following line:

```
GPIO.setup("GPIO0_26", GPIO.OUT)
```

The logic level of an output pin is set with the `output` function, which accepts a pin string and the logic level. If the logic level is set to `GPIO.HIGH` or 1, the pin's voltage is set to 3.3 V. If the logic level is set to `GPIO.LOW` or 0, the voltage is set to 0 V.

The logic level of an input pin is read with the `input` function, whose only argument is the pin string. In my tests, `input` returns 1 if the pin's voltage is greater than 1.4 V and returns 0 if the voltage is less than 1.1 V. If the voltage is between 1.1 V and 1.4 V, `input`'s return value can't be determined.

The code in Listing 11.1 shows how `setup`, `input`, and `output` are used in practice. This script reads from Pin 16 in Header 8, and depending on its value, sets the value of Pin 18.

Listing 11.1 Ch11/test_input.py—Checking a Pin's Logic Level

```
"""
This script repeatedly checks a pin's logic level.
If the logic level is low, a second pin is set high.
If the logic level is high, the loop terminates.
"""


import Adafruit_BBIO.GPIO as GPIO

# Assign names
input_pin = "P8_16";
output_pin = "P8_18";

# Set pin directions
```

```
GPIO.setup(input_pin, GPIO.IN)
GPIO.setup(output_pin, GPIO.OUT)

# Wait for input_pin to reach low voltage
while(GPIO.input(input_pin) == GPIO.LOW):
  GPIO.output(output_pin, GPIO.HIGH)

# Return pins to default state
GPIO.cleanup()
```

In this script, every iteration of the `while` loop checks the state of the input pin. If the pin's level is low, the voltage of the output pin is set high. The loop continues until the input pin's voltage is set high. Then the `cleanup` function returns the pins to their original states.

Events and Event Handling

Many GPIO applications are reactive in nature. That is, they perform operations only in response to external stimuli. In Listing 11.1, the code uses a `while` loop to wait until the input pin's logic level is set high. The Adafruit_BBIO module provides an easier, more flexible way to do this with the `wait_for_edge` function. Its signature is given as follows:

```
wait_for_edge(string pin, int event)
```

When this function executes, it waits until the specified event occurs on the given pin. A GPIO event corresponds to a change in the pin's logic level, which may be a rising edge (low to high) or a falling edge (high to low). The second parameter of `wait_for_edge` specifies the type of event and can take any of the following values:

- `GPIO.RISING`—The function stops waiting when the pin's level changes from low to high.

- `GPIO.FALLING`—The function stops waiting when the pin's level changes from high to low.

- `GPIO.BOTH`—The function stops waiting whenever the pin's logic level changes.

For example, the following code waits until a falling edge occurs on Pin P8_18:

```
wait_for_edge("P8_18", GPIO.FALLING)
```

Rather than halt the processor until an event occurs, it's usually more efficient to check the pin's state periodically while performing other work. This is made possible by the `add_event_detect` and `event_detected` functions. The first is similar to `wait_for_edge`, but instead of halting the processor until the event occurs, it tells the processor to turn on detection for that event.

If detection has been turned on for an event, `event_detected` returns 1 if the event has occurred and 0 if it hasn't. The following code shows how `add_event_detect` and `event_detected` work together to respond when a falling edge takes place on Pin GPIO1_23:

```
add_event_detect("GPIO1_23", GPIO.FALLING)
while(condition == True):
  ...perform other tasks...
  if(event_detected("GPIO1_23")):
    ...respond to the falling edge...
```

It's important to keep in mind that neither `add_event_detect` nor `event_detected` halt the processor. Therefore, this code allows the processor to keep busy as it waits for the event.

The `add_event_detect` function has two optional parameters that affect how the event is processed. Its full signature is given as follows:

```
add_event_detect(string pin, int event, callback=func, bouncetime=time)
```

The third parameter identifies a function, called a callback function, to be called when the event occurs. When called, this function receives a string parameter that identifies the name of the pin that produced the event.

When a user presses a button connected to a GPIO pin, the pin may generate multiple events in rapid succession. Rather than respond to every event, it's more efficient to respond to the first and ignore further events for a given time. The last parameter of `add_event_detect` makes this possible, and it accepts a value for `bouncetime` in milliseconds.

The code in Listing 11.2 shows how `add_event_detect` can be used to execute a callback function. In this program, the `event_callback` function receives the pin string and prints a message.

Listing 11.2   Ch11/callback.py—Responding to Events in a Callback

```
"""
This code configures a callback that responds
to changes to the logic level for Pin P8_18.
"""

import Adafruit_BBIO.GPIO as GPIO
import time

def event_callback(pin):
  print("The event was received by Pin %s." % pin)

# Define pin to be tested
test_pin = "P8_18";

# Set pin direction
GPIO.setup(test_pin, GPIO.IN)

# Configure a callback to be executed
GPIO.add_event_detect(test_pin, GPIO.BOTH, event_callback)
```

```
# Delay for ten seconds
time.sleep(10)

# Return pin to default state
GPIO.cleanup()
```

If the logic level on P8_18 changes, the processor executes the `event_callback` function, passing it the name of the pin. When the function executes, it receives the parameter and prints a message:

```
The event was received by Pin P8_18.
```

The call to `time.sleep` tells the processor to wait 10 seconds before continuing. While the delay continues, the processor continues to check for rising/falling edges on the test pin.

11.3 PWM Generation

As discussed in Chapter 3, "DC Motors," a PWM signal consists of a series of pulses with varying widths. These signals are used to control brushed and brushless DC motors. This section explains how the Adafruit_BBIO.PWM module makes it possible to generate PWM pulses. Table 11.3 lists five of the module's functions.

Table 11.3   Functions of the Adafruit_BBIO.PWM Module

Function	Description
`start(string pin, float duty, freq=freq, polarity=pol)`	Used to generate PWM for the given pin with the specified duty cycle and frequency
`set_duty_cycle(string pin, float duty)`	Used to change the duty cycle for the given pin
`set_frequency(string pin, float freq)`	Used to change the pulse frequency for the given pin
`stop(string pin)`	Halts PWM generation for the given pin
`cleanup()`	Returns pins to their initial configuration

The first function, `start`, is the most important. This has two required parameters: the name of the pin and the desired duty cycle. The duty cycle must be given as a `float` between 0.0 and 100.0. The following code generates a PWM signal on Pin P8_18 with a duty cycle of 25%:

```
GPIO.start("P8_18", 25)
```

`start`'s optional third parameter sets the PWM frequency. By default, this equals 2000 Hz, which means the time between the rising edges of two adjacent pulses is 1/2000 = 0.5 ms. This is strange, because many servomotors expect a PWM period equal to 20 ms, which corresponds to a frequency of 50 Hz. The following call to start generates a PWM signal with a duty cycle of 10% and a frequency of 50 Hz:

```
GPIO.start("P8_18", 10, 50)
```

start's optional fourth parameter makes it possible to change the signal's polarity. This can be set to 0 or 1. The default value is 0, which keeps the signal low until a pulse brings it high. If the polarity is set to 1, the signal stays high until a pulse brings it low. Figure 11.4 shows the difference between the two signals.

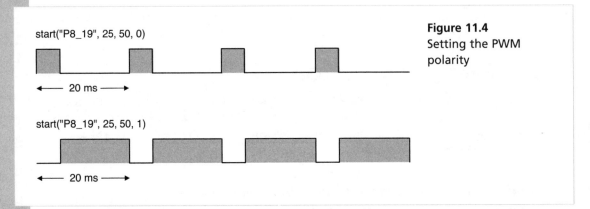

start("P8_19", 25, 50, 0)

← 20 ms →

start("P8_19", 25, 50, 1)

← 20 ms →

Figure 11.4
Setting the PWM polarity

After start is called, set_duty_cycle and set_frequency can be called to change the PWM duty cycle and frequency. As in the start function, the duty cycle and frequency are given as floating-point values.

The code in Listing 11.3 demonstrates how to generate PWM signals using the Adafruit_BBIO.PWM module. The program delivers pulses to Pin P8_19, and each pulse has a duty cycle of 40%. In keeping with servo requirements, the frequency is set to 50 Hz.

Listing 11.3 Ch11/pwm.py—Generating a PWM Signal

```
"""
This code generates a pulse-width modulation (PWM) signal
for Pin P8_19 with a 40% duty cycle and a frequency of 50 Hz.
"""

import Adafruit_BBIO.PWM as PWM
import time

# Define PWM pin
pwm_pin = "P8_19"

# Set duty cycle to 40%, frequency to 50 Hz
PWM.start(pwm_pin, 40, 50)
```

```
# Delay for ten seconds
time.sleep(10)

# Halt PWM and return pin to initial settings
PWM.stop(pwm_pin)
PWM.cleanup()
```

The PWM signal is delivered through P8_19, and this wasn't selected arbitrarily. This pin is connected to the processor's enhanced high-resolution pulse width modulator, or EHRPWM. You can check this by looking at Pages 70 and 72 of the BBB's system reference manual. Other suitable PWM pins are P8_13, P9_14, and P9_16.

Using the Adafruit_BBIO library, programs can generate PWM signals with precise timing. However, the BBB's pins deliver a maximum of 3.3 V, so the amplitude of the pulses can't exceed 3.3 V. This isn't enough voltage to power a servomotor directly, but it's more than sufficient to provide control signals to an electronic speed control (ESC) or an expansion board such as the Dual Motor Controller Cape, which is the topic of the following section.

11.4   The Dual Motor Controller Cape (DMCC)

The BeagleBone Black's capabilities can be extended with expansion boards that connect to its P8 and P9 headers. These expansion boards are referred to as *capes*, presumably because Snoopy the beagle wore a cape when imagining himself as the Red Baron.

BBB capes have been designed for many applications, including audio processing, LCD control, and Wi-Fi communication. The site http://elinux.org/Beagleboard:BeagleBone_Capes presents a wide range of capes that are compatible with the BBB.

The Dual Motor Controller Cape, or DMCC, is sold by Exadler Technologies at http://exadler. myshopify.com. Figure 11.5 shows what it looks like. The headers on the top and bottom make it possible to stack additional DMCC boards.

The three connectors on the right are particularly important. The top and bottom connectors deliver power to two motors, and keeping with Exadler's documentation, we'll refer to them as Motor 1 and Motor 2. The middle connector receives external power. The input voltage must be set between 5 V and 28 V.

The DMCC's processing is performed by three integrated circuits. A DSPIC33FJ32MC304 generates PWM signals to set the motors' speeds. Two VNH5019A devices provide H bridges that receive PWM signals and deliver current to motors in the forward or reverse directions. Figure 11.6 gives a better idea of how the DMCC's components work together.

The full schematic and design files for the DMCC can be downloaded for free at https://github.com/ Exadler/DualMotorControlCape. To read these files, you need to install the EAGLE circuit design tool. For more information on EAGLE, I recommend *Designing Circuit Boards with EAGLE* by Matthew Scarpino.

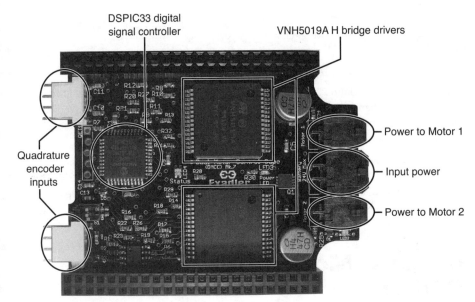

Figure 11.5
The Dual Motor Controller Cape, Mk 6

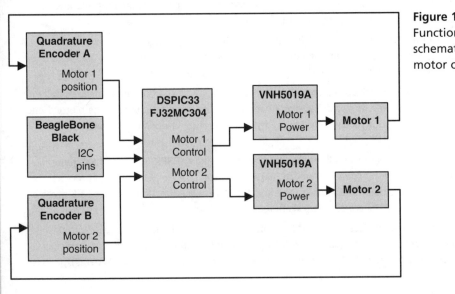

Figure 11.6
Functional
schematic of DMCC
motor control

As illustrated in Figure 11.6, the DMCC's operation consists of four steps:

1. A program running on the BeagleBone Black specifies which motors should rotate and the speed and direction of the rotation.

2. The DSPIC33 receives the desired motor parameters from the BBB and the motors' actual behavior from the quadrature encoders. With this information, it generates PWM signals for the motors.

3. The H bridges in the VNH5019A deliver current to the connected motors.

4. The motors rotate as they receive current. Their quadrature encoders convert their shaft angles into electrical signals.

This section discusses these steps in detail and then shows how to control DC motors with the DMCC.

11.4.1 BBB-DMCC Communication

Despite the many connections in the header, the DMCC reads data from only two BBB pins: P9_19 and P9_20. They tell the DMCC how the motors should rotate.

The data transfer protocol used by these pins is I2C, which stands for Inter-Integrated Circuit. This simple method relies on two signals to carry data:

- **Serial data line (SDA)**—Transfers bits between devices

- **Serial clock (SCL)**—Data clock

In I2C communication, the device that drives the clock is called the master and the other device or devices are called slaves. In BBB-DMCC communication, the BBB is the master and each connected DMCC is a slave. The lowest DMCC is Slave 0, and each DMCC stacked on top of it receives an ID incremented by 1.

Every I2C data transfer consists of a sequence of 8-bit messages. The master initiates a transfer by holding SCL high and changing SDA from high to low. The message starts with a slave's ID and a bit that identifies whether the master intends to send or receive data. The transfer ends when the master holds SCL high and changes SDA from low to high.

11.4.2 PWM Signal Generation

The DMCC controls the motors' speeds using PWM signals generated by the DSPIC33FJ32MC304, which I'll shorten to DSPIC33. This device is a *digital signal controller*, which is a microcontroller with so much number-crunching power that it can serve the same role as a digital signal processor.

In the DMCC, the DSPIC33 reads two pieces of information. From the BBB's I2C connection, it receives the desired motor parameters. From the two quadrature encoder inputs, it reads the current speed of the motors.

The DSPIC33 computes the error between the desired speed of the motors and their actual rotation. To reduce this error, it generates a control signal using the PID (proportional-integral-differential) method discussed in Chapter 5, "Servomotors." This control signal is the sum of three values:

- **Proportional gain**—Proportional to the current error

- **Integral gain**—Proportional to the sum of the error over time

- **Differential gain**—Proportional to the most recent change in error

The DSPIC33 adds these values together to generate the PWM signals that need to be sent to the motors. More precisely, it determines the PWM duty cycle that will reduce the difference between the motors' desired behavior and their actual behavior. Then the device delivers PWM pulses and enable signals to the VNH5019A connected to the desired motor.

11.4.3 Switching Circuitry

The DMCC contains two VNH5019A devices, and each VNH5019A contains two half-H bridges. As described in Chapter 3, an H bridge makes it possible to drive a motor in forward or reverse by delivering positive or negative current, respectively. Figure 11.7 gives an idea of how this works.

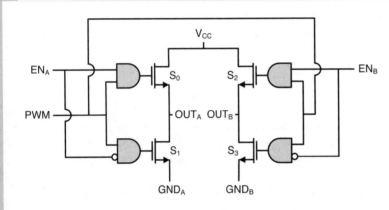

Figure 11.7
Operation of the VNH5019A H bridge

The two half-H bridges in the VNH5019A are denoted as A and B. Each has separate enable signals (EN_A and EN_B) and delivers a separate output (OUT_A and OUT_B). In general, OUT_A is connected to one terminal of a motor and OUT_B is connected to the other.

There's only one PWM signal. When this is high, the switches open and close according to the values of EN_A and EN_B. For example, when PWM and EN_A are high, OUT_A receives current from V_{CC}. When PWM is low, the switches remain open, so no current flows to OUT_A or OUT_B.

In the DMCC, the values of EN_A, EN_B, and PWM are set by the DSPIC33. It's safe to assume that EN_B is the inverse of EN_A. If this is the case, switches S_0 and S_3 will be closed when EN_A is high, driving current from OUT_A to OUT_B. When EN_A is low, switches S_1 and S_2 will be closed, driving current from OUT_B to OUT_A.

11.4.4 Motor Control

Exadler Technologies has provided source code for controlling motors with the DMCC. The files can be freely downloaded from https://github.com/Exadler/DMCC_Library. Once you've downloaded the files to the BBB, change to the directory containing setupDMCC.py and enter the following command:

```
python setupDMCC.py install
```

This compiles the *.c code and installs the DMCC module in Python's dist-packages directory. Once this is done, you can write and execute scripts that make use of the DMCC module. Table 11.4 lists ten of the module's functions.

Table 11.4 Functions of the DMCC Module

Function	Description
getMotorCurrent(int board, int motor)	Returns the current of the given motor
getMotorDir(int board, int motor)	Returns the direction of the given motor
getMotorVoltage(int board, int motor)	Returns the voltage of the given motor
getQEI(int board, int motor)	Returns the quadrature encoder value for the given motor
getQEIDir(int board, int motor)	Returns the quadrature encoder direction for the given motor
getTargetPos(int board, int motor)	Returns the position of the given motor
getTargetVel(int board, int motor)	Returns the velocity of the given motor
setMotor(int board, int motor, int power)	Delivers the specified power to the given motor
setPIDConstants(int board, int motor, int posOrVel, float P, float I, float D)	Assigns the PID values to control the given motor's position or velocity
setTargetPos(int board, int motor, int pos)	Sets the position of the given motor

In each of these functions, the first parameter identifies the board of interest. This is useful if multiple DMCC boards are stacked on top of one another. The bottom board is specified with 0 and successive boards have values incremented by 1.

The second parameter sets the motor of interest. This should be set to 1 to specify Motor 1 and 2 to specify Motor 2.

Of these functions, the most important is setMotor, which drives a motor with the specified amount of power. The power value can be set to any integer between –10000 and 10000. A positive value drives the motor in the forward direction, and a negative value drives the motor in the reverse direction.

For example, the following code drives Motor 1 on Board 0 in the forward direction with power set to half of the maximum:

```
setMotor(0, 1, 5000)
```

Listing 11.4 presents a simple script that drives the motor forward at full speed for 5 seconds and then stops. Then it drives the motor in reverse at half speed for 10 seconds and stops.

Listing 11.4   Ch11/motor.py—Controlling a Motor with the DMCC

```
"""
This drives a motor forward at full speed for 5 seconds, stops,
drives the motor backward at half-speed, and stops.
"""

import DMCC as DMCC
import time

# Drive motor forward
setMotor(0, 1, 10000)
time.sleep(5)

# Stop
setMotor(0, 1, 0)
time.sleep(3)

# Drive motor backward
setMotor(0, 1, -5000)
time.sleep(10)

# Stop
setMotor(0, 1, 0)
time.sleep(3)
```

The DMCC module is easy to work with, but one shortcoming is the difficulty in driving a stepper motor. With two H bridges available, it isn't difficult to code a Python function that delivers the phase changes required by a stepper. However, in the DMCC module, the only motor-driving function is setMotor, which doesn't allow you to configure the output signals.

11.5 Summary

The BeagleBone Black is one of the most powerful single-board computers available for hobbyists. It lacks a video decoder, but its ARM processor can process data and crunch numbers at high speed. Its many external connectors make it possible to access the board in a number of ways.

This chapter has focused on programming in Python on the Debian OS, but these aren't the only options available. A number of Linux-based operating systems can run on the BBB, including Ubuntu and Ångström. Applications can be written in many languages, including C, C++, Java, and BoneScript.

If you want to access the BBB's pins in Python, you can't do much better than the Adafruit_BBIO library. This library provides the Adafruit_BBIO.GPIO module, whose functions read and set the voltage levels of GPIO pins. It also provides functions that wait for rising and falling edges to occur on GPIO pins.

Adafruit_BBIO also provides the Adafruit_BBIO.PWM module, whose functions generate pulse width modulation (PWM) signals. Keep in mind that the default PWM frequency is 2000 Hz, which is much faster than the 50 Hz expected by many hobbyist servomotors. Also, the maximum PWM amplitude is 3.3 V, which is less than the required voltage for many servomotors.

The Dual Motor Controller Cape (DMCC) is an extension board for the BBB that makes it possible to control motors. A digital signal controller reads the motors' positions and constructs a control signal based on the PID control method. It delivers PWM pulses to a pair of H bridge components, and these components deliver power to the motors. The functions provided by the DMCC module make it possible to configure the motors' behavior and read their operating characteristics.

12

DESIGNING AN ARDUINO-BASED ELECTRONIC SPEED CONTROL (ESC)

Chapters 9 through 11 explained how to control motors using popular boards such as the Arduino Mega and Raspberry Pi. This chapter puts aside existing boards and explains how to design a motor control board from scratch. To be specific, the goal is to design an electronic speed control (ESC) capable of controlling a brushless DC motor (BLDC). As discussed in Chapter 3, "DC Motors," an ESC receives low-voltage signals from a controller and generates the high-current pulses needed to drive the motor.

For the sake of simplicity, this chapter's ESC will be constructed as an extension board for Arduino family of circuit boards (particularly the Arduino Mega). This means the extension board will receive signals from the Arduino's Atmel microcontrollor and deliver power to a BLDC.

This chapter goes into great depth on a number of topics related to BLDC control, including the electrical characteristics of MOSFETs and MOSFET drivers. It also presents the process of designing the circuit board, from connecting components in the schematic to positioning device packages in the board design. However, before getting into technical detail, the first section presents an overview of the circuit's design process.

12.1 Overview of the ESC Design

For this chapter, I'd originally intended to present a PC-programmable ESC similar to a commercial ESC for RC cars and aircraft. That is, the circuit receives power and PWM pulses, and it delivers control signals to the BLDC. To ensure proper timing, the ESC also receives information from the motor about the rotor's orientation. Figure 12.1 illustrates the overall operation.

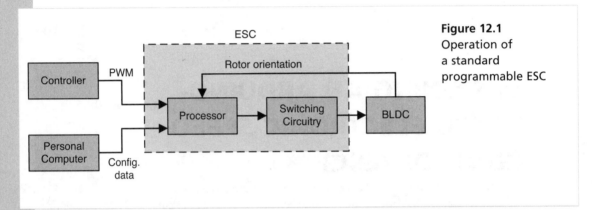

Figure 12.1
Operation of a standard programmable ESC

This type of circuit is popular in the RC community, but as a maker project, it has three significant issues:

- To use the circuit, an external controller is needed to generate PWM signals.

- Programming the ESC's microcontroller requires a deep understanding of the C programming language and the microcontroller's architecture.

- To make the ESC programmable, additional software must be written to run on a personal computer.

Given the complexity of testing and programming a standard ESC, I decided to combine the ESC's functionality with the simplicity of the Arduino framework. Therefore, the ESC circuit presented in this chapter receives control signals from an Atmel microcontroller on a standard Arduino board.

Figure 12.2 depicts the functional schematic for the simplified circuit. I'd considered giving it a name such as the ArduESC or ESCuino, but in the interest of good taste, this chapter will refer to it simply as the ESC Shield.

Figure 12.2
Functional schematic of the
ESC Shield

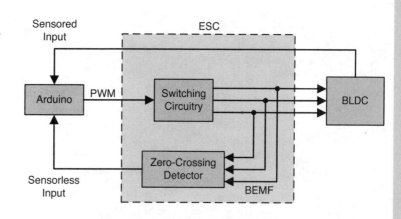

This schematic's structure is reflected in the design of the circuit board. The switching is performed by six transistors, and Figure 12.3 shows how they're positioned on the front side of the circuit board described in this chapter.

Figure 12.3
The front side of the ESC Shield

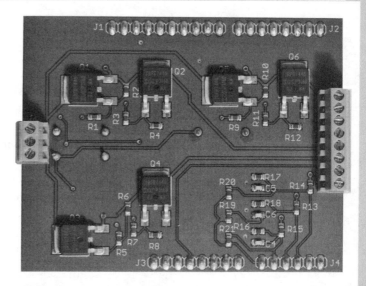

Figure 12.4 shows what the rear side of the board looks like. The pins on the top and bottom fit into the headers on an Arduino board.

The next two sections present the components that make up the ESC's functional blocks. The next section discusses the board's switching circuit, and the section after that explains the zero-crossing circuit.

Figure 12.4
The rear side of the ESC Shield

12.2 Switching Circuitry

DC motors require more power than microcontrollers can provide. For this reason, motor control circuits place electric switches between the controller and the motor. These switches receive signals from the controller and deliver power to the motor in pulses of high voltage and current.

As discussed in Chapter 3, the two most common types of electrical switches are metal-oxide-semiconductor field-effect transistors (MOSFETS) and insulated-gate bipolar transistors (IGBTs). MOSFETs are better suited to small- and medium-sized circuits, so the ESC Shield relies on MOSFETs to switch motor power on and off. The first part of this section goes into detail regarding which MOSFETs were chosen and why.

To ensure that the MOSFETs operate at maximum speed, the ESC circuit places additional devices between the Arduino microcontroller and the MOSFETs. These are called MOSFET drivers, and the second part of this section explains how they work.

Figure 12.5 expands on Figure 12.2 and shows how the circuit's MOSFETs and MOSFET drivers are connected to one another.

12.2.1 MOSFET Switches

Chapter 3 explained the basics of MOSFETs, and I'll briefly review the topic here. Figure 12.6 depicts the schematic symbol of an n-type enhancement-mode MOSFET. As shown, the device has three terminals: the gate, source, and drain.

Figure 12.5
ESC circuit with
MOSFET drivers
and MOSFET
switches

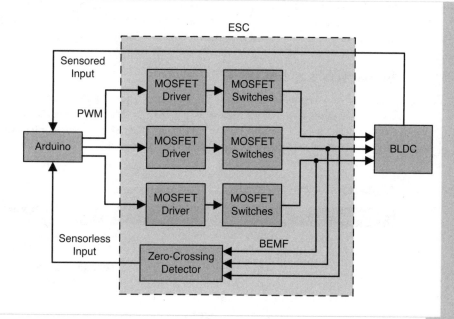

Figure 12.6
The MOSFET circuit symbol

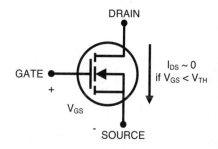

Enhancement-mode MOSFETs can operate in three states, but for motor circuits, we're only concerned with two:

- **The cut-off state (off)**—If the gate-source voltage (V_{GS}) is lower than a threshold voltage (V_{TH}), the MOSFET is in the cut-off state. The resistance between the source and drain is nearly infinite and the drain-source current, I_{DS}, is nearly zero.

- **The saturation state (on)**—If V_{GS} is greater than V_{TH} and the drain-source voltage (V_{DS}) is greater than V_{GS} - V_{TH}, the MOSFET is in the saturation state. The resistance between the drain and source, $R_{DS(ON)}$, drops to a few hundredths of an ohm. This allows current to flow freely from the drain to the source.

MOSFETs are not ideal switches. The threshold voltage is greater than 0 V, the drain-source resistance is greater than 0Ω, and the switching time isn't instantaneous. However, their behavior is sufficiently close to the ideal that most motor circuits you'll encounter rely on them for switching.

Selecting a MOSFET

If you look at a datasheet for a MOSFET, you'll find a vast list of parameters that describe the transistor's operation. With names such as $V_{BR(DSS)}$ and Q_G, it's understandable why newcomers are frequently confused. To help clear the confusion, Table 12.1 lists six parameters commonly encountered in MOSFET datasheets and explains what they mean.

Table 12.1 MOSFET Operational Parameters

Parameter	Full Name	Description
V_{DS} or $V_{BR(DSS)}$	Drain-source voltage	Maximum voltage between drain and source that the MOSFET can block in its off state
I_D	Drain current	Maximum current that can flow between the drain and source when the MOSFET is in the on state
$R_{DS(ON)}$	Drain-source resistance	Resistance between the drain and source when the MOSFET is in the on state
Q_G	Gate charge	The amount of charge needed at the gate to put the MOSFET in the on state
P_d or P_{TOT}	Total power	Maximum amount of power that can be dissipated by the MOSFET
$V_{GS(TH)}$ or V_{TH}	Threshold voltage	Gate-source voltage needed to put the MOSFET in the on state

The first two characteristics are particularly important. If a MOSFET can't withstand the power supply's voltage or conduct the amount of current required by the motor, it will fail and potentially break the circuit.

Motors require significant amounts of current, so the resistance, $R_{DS(ON)}$, is a major concern. The larger the resistance, the larger the voltage drop between the drain and source. High resistance means more heat, which equals $I^2R_{DS(ON)}$. I prefer MOSFETs with resistances around 5 mΩ, so even if the current is 20 A, the voltage drop is only 0.1 V. For high-power motor control circuits, IGBTs are preferred over MOSFETs because the voltage drop is smaller.

The gate charge, Q_G, is also important. A MOSFET's gate voltage, V_{GS}, doesn't change instantaneously. Similar to a capacitor, its voltage depends on the charge it receives (voltage = charge/capacitance). Q_G identifies the amount of charge the gate needs to reach the threshold voltage. It takes time to charge and discharge the gate, so the smaller the Q_G, the faster the gate's voltage will reach the threshold voltage for a given amount of current.

The IRFR7446 MOSFET

The IRFR7446 power MOSFET from International Rectifier delivers high power with low resistance and low gate charge. For this reason, I've selected this device to serve as the switch in the ESC Shield. Its operating characteristics are given as follows:

- **Maximum drain-source voltage ($V_{BR(DSS)}$):** 40 V

- **Maximum drain-source current:** 120 A

- **On-state drain-source resistance ($R_{DS(ON)}$):** 3.9 mΩ

- **Typical gate charge (Q_G):** 65 nC at V_{GS} = 10 V

The gate charge of 65 nC is low compared to most power MOSFETs, but for high-speed power switching, this charge must be provided in a very short amount of time. This means the MOSFET's gate must be driven at high current. The Arduino microcontroller can't provide the required level of current, so the ESC makes use of MOSFET drivers. I'll discuss them shortly.

Body Diodes

Regarding MOSFETs, one more topic needs to be discussed. In a motor control circuit, MOSFETs are commonly connected in pairs, such as that shown in Figure 12.7.

Figure 12.7
MOSFETs and diodes

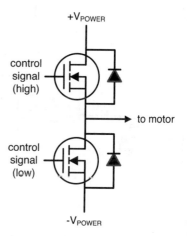

In this figure, each MOSFET has a diode connected in parallel that conducts current from the source to the drain. These diodes provide a path for current to flow away from the motor's windings after the MOSFET has turned off. These diodes are frequently called free-wheeling diodes, suppression diodes, or flyback diodes.

In many MOSFET circuits, Schottky diodes are commonly used because of their rapid switching and low forward voltage drop. However, power MOSFETs, such as the IRFR7446, have built-in diodes called *body diodes*. Datasheets for power MOSFETs provide two important operating characteristics related to body diodes. In both cases, the lower the value, the better.

- V_{SD}—The forward voltage drop of the body diode (0.9 V for the IRFR7446)

- t_{rr}—The reverse recovery time (20 ns for the IRFR7446)

This second parameter merits explanation. When a diode's voltage is switched from forward to reverse, its stored energy causes current to flow briefly in the reverse direction. This brief time is called the *reverse recovery time*, and in motor circuits, it's important that this be as small as possible.

12.2.2 MOSFET Drivers

For a MOSFET to switch power quickly, its gate must be rapidly charged and discharged. This requires more current than the Arduino's microcontroller can provide by itself. For this reason, high-speed ESCs insert amplifiers between the microcontroller and the MOSFETs. These devices are referred to as MOSFET drivers or charge pumps.

Some circuits use one driver per MOSFET, but it's easier to use an IC capable of driving the two MOSFETs in a half-H bridge. Figure 12.8 gives an idea of how this works.

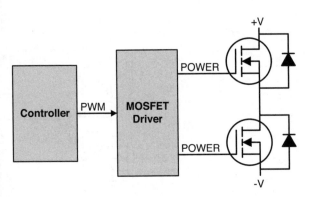

Figure 12.8
MOSFET driver for a half-H bridge

The MOSFET that connects the motor to positive power is the *high-side* transistor and the MOSFET that connects the motor to negative power is the *low-side* transistor. A driver that can provide switching current to both is called a high-side/low-side driver.

When you're selecting the right driver, the first step is to determine how much current is needed by the switch. This depends on the MOSFET's gate charge and how quickly the MOSFET needs to switch power on and off. Mathematically, the relationship between the gate charge (Q_G), the time needed to switch on/off (t_{switch}), and the required current (I) is given as follows:

$$I = \frac{Q_G}{t_{switch}}$$

For the IRFR7446, Q_G equals 65 nC when V_{GS} = 10 V. For the ESC, a safe switching time is 500 ns. Replacing these values in the equation produces the following result:

$$I = \frac{65nC}{500ns} = 0.13A$$

The IR2110 High-Side/Low-Side Driver

To drive its MOSFETs, the ESC Shield relies on the IR2110 high-side/low-side driver. This driver's typical output current is 2 A, which is more than sufficient to drive the IRFR7446 MOSFET. The surface-mount package has 16 pins, but only 11 of them are used. Figure 12.9 shows how these 11 pins are connected to the controller, to power, and to the high-side/low-side MOSFETs.

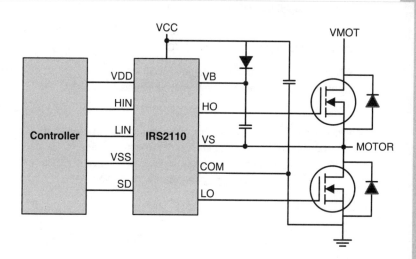

Figure 12.9
The IR2110 high-side/low-side driver circuit

The IR2110 signals are given as follows:

■ HIN and LIN receive PWM signals from the controller.

■ VDD and VSS are the voltage and ground as seen from the controller.

■ HO and LO provide switching power to the MOSFETs' gate terminals.

■ VB and VS serve as the high-side and low-side floating supplies.

■ VCC is the low-side fixed supply voltage.

■ COM is the low-side return (ground).

■ SD tells the IR2110 to turn off (shut down) the output voltages HO and LO.

The driver's overall operation is straightforward to understand. When HIN is high, HO switches power to turn the high-side MOSFET fully on. When LIN is high, LO switches power to turn the low-side MOSFET fully on. It is vital that HIN and LIN are never high at the same time. If they're both high, the result is a short circuit from power to ground. This is called *shoot through*, and it can cause significant overheating problems.

The different voltage inputs can be confusing. For the ESC Shield, VDD is set to 5 V. VCC is set between 10 and 20 V, and the ESC Shield sets it to 12 V. If VCC reaches a voltage less than about 8.5 V, the IR2110's undervoltage detector prevents HO and LO from delivering current.

12.2.3 Bootstrap Capacitor

To keep the high-side MOSFET fully on, its gate voltage must be raised 10 to 15 V above the source voltage. However, the source voltage can be as high as VMOT, which is higher than VCC. To provide the IRS2110 with enough power to drive the high-side MOSFET, many circuits connect a capacitor across VB and VS. This is called a bootstrap capacitor.

To see why this capacitor is helpful, it's important to examine the circuit's behavior as the high-side and low-side MOSFETs switch on and off.

- When LIN is high and the low-side MOSFET is fully on, VS is connected to ground. In this case, the bootstrap capacitor charges to VCC minus the diode drop.

- When HIN is high and the high-side MOSFET is fully on, VS is connected to VMOT. Now the potential difference across the bootstrap capacitor (VB – VS) equals VMOT plus VCC minus the diode drop.

Choosing the capacitance of the bootstrap capacitor requires effort. If the capacitance is too low, it won't be able to hold enough charge to raise the voltage of VB. If the capacitance is too high, it will take too long to charge properly. As given by Application Note 978 (AN978) from International Rectifier, the capacitance can be computed with the following equation:

$$C \geq \frac{2\left[2Q_G + \dfrac{I_{qbs(\max)}}{f} + Q_{ls} + \dfrac{I_{Cbs(\text{leak})}}{f}\right]}{V_{cc} - V_f - V_{LS} - V_{Min}}$$

Table 12.2 provides a description of the variables in this expression. The third column lists the approximate values for the components in the ESC Shield.

Table 12.2 Variables in the Bootstrap Capacitance Equation

Variable	Description	Value
Q_G	Gate charge of the high-side MOSFET	65 nC
$I_{qbs(\max)}$	Maximum V_{BS} quiescent current	230 μA
f	Operating frequency	50 Hz
Q_{ls}	Level shift charge required per cycle	5 nC

Variable	Description	Value
$I_{Cbs(leak)}$	Capacitor leakage current	0.5 µA
V_{CC}	Supply voltage	12 V
V_f	Voltage drop across bootstrap diode	0.7 V
V_{LS}	Voltage drop across low-side MOSFET	0.06 V
V_{Min}	Minimum difference between V_B and V_S	9.4 V

In this table, the voltage drop across the low-side MOSFET is obtained by multiplying the desired current by the MOSFET's $R_{DS(on)}$ value. For the ESC Shield, this is (20 A)(3.0 mΩ) = 0.06 V. Also, because tantalum capacitors have nearly no leakage current, this value may be set to zero.

When the table's values are inserted into the equation, the result is a capacitance of about 5.16 µF. For this reason, the ESC Shield will use a 4.7 µF tantalum capacitor to serve as the bootstrap capacitor.

12.3 Zero-Crossing Detection

In my opinion, the main disadvantage of working with BLDCs is that the controller needs to determine the motor's state in order to deliver power. This is straightforward when the motor has a sensor or encoder, but if no sensor or encoder is available, the circuit becomes more complicated.

Chapter 3 explained the basic process of controlling a three-phase BLDC—the controller drives two phases at a time, leaving the third to float. By measuring the floating voltage of the undriven phase, the controller can determine when to send power to a motor. This is called zero-crossing detection, and it's the most common method for determining when to deliver power to a sensorless motor.

As the motor rotates, it generates an internal voltage in its windings called the back-EMF. Figure 12.10 shows a basic approximation of the back-EMF in a BLDC. In a real circuit, the shape of a winding's back-EMF can be dramatically different when the winding isn't receiving voltage.

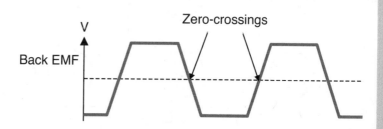

Figure 12.10
Ideal back-EMF of a three-phase BLDC

The point at which a winding's back-EMF crosses zero is difficult to obtain because a motor's back-EMF can't be measured directly. To see why this is the case, consider the circuit in Figure 12.11. This contains the equivalent circuit for the BLDC and a portion of the control circuitry.

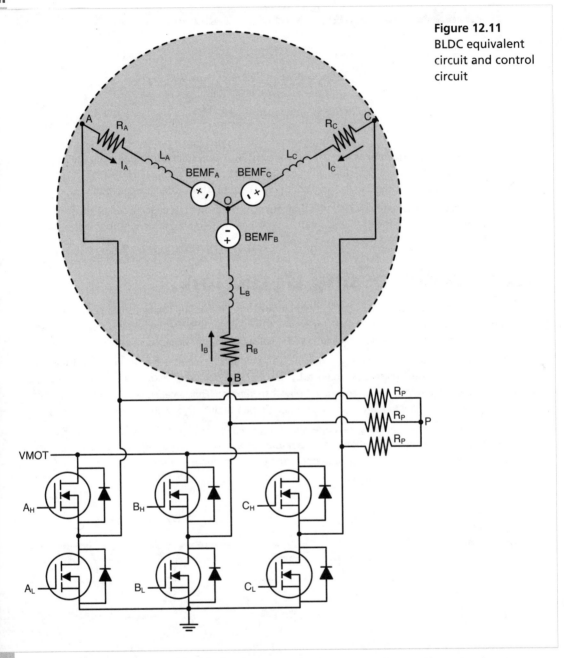

Figure 12.11
BLDC equivalent circuit and control circuit

In this diagram, the BLDC's three windings are connected at the point O. This point is commonly called the neutral point or the star point. If we knew the voltage at O, it would be straightforward to determine when the back-EMF in the floating winding crosses zero volts.

Unfortunately, we can't obtain V_O without drilling into the motor, so we create a virtual neutral point at point P outside the motor. This allows us to find the zero-crossing by measuring V_P and the voltage of the floating winding.

Four steps are required to derive the relationships between the zero-crossing, V_P, and the floating voltage:

1. Relate V_P to the voltages of the three windings: V_A, V_B, and V_C.

2. Relate V_O to the voltages of two energized windings.

3. Relate V_O to the voltage of the floating winding and the floating back-EMF.

4. Combine the results to solve for the floating back-EMF.

The remainder of this section presents these steps in the given order.

12.3.1 Step 1: Relate V_P to the Voltages of the Three Windings

The first step is simple. The virtual neutral point, P, connects to each winding through a resistor whose resistance equals R_P. The voltage at P, V_P, can be computed in terms of the winding voltages (V_A, V_B, and V_C) using Kirchoff's Current Law:

$$\frac{V_A - V_P}{R_P} + \frac{V_B - V_P}{R_P} + \frac{V_C - V_P}{R_P} = 0$$

$$V_A + V_B + V_C - 3V_P = 0$$

$$V_P = \frac{V_A + V_B + V_C}{3}$$

12.3.2 Step 2: Relate V_O to the Voltages of Two Energized Windings

In the figure, suppose that switches B_H and C_L are closed and the rest of the switches are open. This means Winding B is connected to high voltage ($V_B = V_{HIGH}$) and Winding C is connected to low voltage ($V_B = 0$). Because the other switches are open, no current flows to Winding A, whose voltage is left floating. With $I_A = 0$, Kirchoff's Current Law at point O tells us that $I_C = -I_B$.

An energized winding generates back-EMF in response to applied current. Windings B and C have currents of equal magnitude but opposite direction, so their back-EMF voltages have equal magnitude and opposite sign. In other words, because $I_C = -I_B$, $BEMF_C = -BEMF_B$.

Now let's compare the voltage across Winding B ($V_B - V_O$) and the voltage across Winding C ($V_C - V_O$). These potential differences can be computed with the following equations:

$$V_B - V_O = I_B R_B + L_B \frac{dI_B}{dt} + BEMF_B$$

$$V_C - V_O = I_C R_C + L_C \frac{dI_C}{dt} + BEMF_C$$

It's safe to assume that the resistances and inductances are the same for both windings. Replacing I_B with $-I_C$ and $BEMF_B$ with $-BEMF_C$, the equation for the voltage across Winding B can be given as follows:

$$V_B - V_O = -I_C R_C - L_C \frac{dI_C}{dt} - BEMF_C$$

$$= -(V_C - V_O)$$

Rearranging this relationship produces the following result:

$$V_O = \frac{V_B + V_C}{2}$$

12.3.3 Step 3: Relate V_O to the Voltage of the Floating Winding and the Floating Back-EMF

If switches A_H and A_L are open, no current flows through Winding A. This means there's no voltage drop across R_A or L_A. Therefore, the voltage across Winding A can be computed with the following equation:

$$V_A - V_O = BEMF_A$$

Rearranging this equation produces the following expression for V_O:

$$V_O = V_A - BEMF_A$$

12.3.4 Step 4: Combine the Results to Solve for the Floating Back-EMF

At this point, we have three equations. Two solve for V_O and one solves for V_P:

$$V_O = \frac{V_B + V_C}{2}$$

$$V_O = V_A - BEMF_A$$

$$V_P = \frac{V_A + V_B + V_C}{3}$$

By manipulating the first two equations, we can arrive at the following equation for the floating back-EMF:

$$BEMF_A = \frac{2V_A - V_B - V_C}{2}$$

Combining this result with the third equation produces the final result:

$$V_P - V_A = -\frac{2}{3} BEMF_A$$

The exact value of the floating back-EMF isn't important. What is important is that it crosses zero at the same moment that $V_P - V_A$ equals zero. Similarly, the floating back-EMF of Winding B equals zero when $V_P - V_B$ equals zero, and the floating back-EMF of Winding C equals zero when $V_P - V_C$ equals zero. By measuring $V_P - V_A$, $V_P - V_B$, and $V_P - V_C$, the controller can detect the zero-crossing point of each winding in a straightforward manner.

12.4 Designing the Schematic

Like the Arduino Mega and the Arduino Motor Shield, the ESC Shield is designed with the EAGLE circuit design tool. This design process consists of two steps: constructing the circuit's schematic with the component's symbols and then laying out the actual circuit with the component's packages.

This section focuses on the circuit's schematic design, which is performed in EAGLE's schematic editor. The goal of the schematic is to identify the circuit's components and the manner in which they're connected. Rather than connect all the components in a single mess of wires, I like to split the design into subcircuits. The schematic for the ESC Shield can be split into three main subcircuits:

- Header connections

- MOSFETs and MOSFET drivers

- Zero-crossing detection

This section discusses each subcircuit and shows what it looks like in the schematic. If you have EAGLE, you can view the schematic by downloading the book's archive from http://www.motorsformakers.com. The esc_shield.sch file can be found in the Ch12 folder.

12.4.1 Header Connections

One of the many advantages of the Arduino framework is that its circuit boards have identical connections for expansion boards. That is, Arduino boards have header connections with the same sizes and positions.

To see what I mean, look back at Chapter 9, "Motor Control with the Arduino Mega," and compare the headers of the Arduino Mega and the Arduino Motor Shield. The motor shield has four headers: one with 10 pins, two with eight pins, and one with six pins. This arrangement is standard among

Arduino shields, so you'll find the same four headers on the Arduino Proto Shield and the Arduino LCD Shield.

The ESC Shield relies on similar headers to connect to an Arduino board. In addition, it has two headers that connect to the BLDC and a power supply. Figure 12.12 illustrates the four Arduino headers (J1-J4), the BLDC header (J5), and the power supply header (J6).

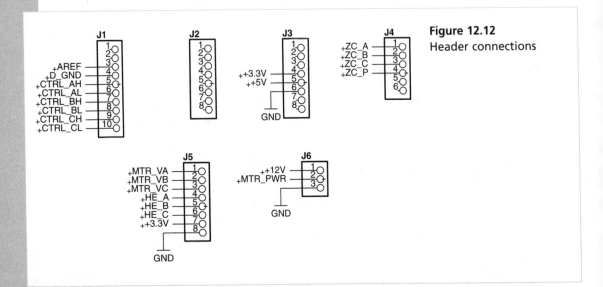

Figure 12.12
Header connections

This figure illustrates the signal names used throughout the ESC Shield. Table 12.3 lists these signals and their headers.

Table 12.3 ESC Shield Signals

Signal	Description
CTRL_AH	High control signal for Winding A
CTRL_AL	Low control signal for Winding A
CTRL_BH	High control signal for Winding B
CTRL_BL	Low control signal for Winding B
CTRL_CH	High control signal for Winding C
CTRL_CL	Low control signal for Winding C
ZC_A	Check zero-crossing for Winding A
ZC_B	Check zero-crossing for Winding B
ZC_C	Check zero-crossing for Winding C
ZC_P	Voltage of the virtual neutral point

Signal	Description
HE_A	Hall effect signal for Winding A
HE_B	Hall effect signal for Winding B
HE_C	Hall effect signal for Winding C

The following subcircuits show how these signals are connected components inside the ESC Shield.

12.4.2 MOSFETs and MOSFET Drivers

The ESC Shield uses MOSFETs to switch power to the BLDC and MOSFET drivers to deliver power to them. Each winding in the BLDC requires a pair of MOSFETs, and each pair is controlled by a single MOSFET driver. Therefore, the ESC Shield contains three MOSFET drivers and six MOSFETs. Figure 12.13 shows how they're connected in the circuit.

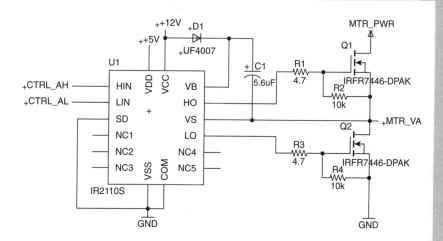

Figure 12.13
MOSFET driver subcircuit

In this figure, C1 is the bootstrap capacitor that raises the voltage at VB. As discussed earlier, its capacitance is 4.7 µF.

This subcircuit places two 10 kΩ resistors (R2 and R4) between the gate and source of each MOSFET. This pulls down the transistors' gates, which helps prevent the MOSFETs from being turned on by external voltage sources, such as static electricity.

The circuit also places two 4.7 Ω resistors in series with the gate of each MOSFET. This slightly reduces the switching efficiency (–0.8%) but significantly decreases the ringing in the circuit, thus increasing stability. For more information, I recommend the AB-9 application bulletin from Fairchild Semiconductor.

12.4.3 Zero-Crossing Detection

An earlier discussion explained how a winding's zero-crossing can be detected by subtracting the winding's voltage from the voltage of a virtual neutral point, P. Figure 12.14 shows the subcircuit that accomplishes this.

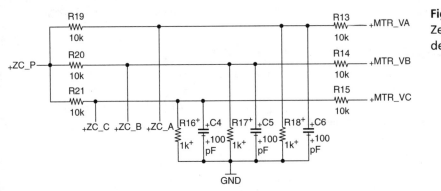

Figure 12.14
Zero-crossing detection subcircuit

The Arduino's analog inputs can only read voltages between 0 and 5 volts. The motor voltage can be considerably larger, so a voltage divider is needed to reduce the winding voltages. In each case, a capacitor is inserted in parallel to filter ringing from the motor voltage.

12.5 Board Layout

With the schematic design completed, the next step is to design the circuit board. In keeping with the convention established for Arduino shields, its dimensions are 2.1" by 2.7", with the four Arduino headers (J1-J4) positioned on the long edges. The J5 header, which connects to the motor, is placed on one short edge. The J6 header, which connects to external power, is placed on the other short edge.

The shield isn't large enough to contain all six MOSFETs and all three drivers on the same side. Therefore, I've positioned the MOSFETs on the front side and the drivers on the rear. Figure 12.15 shows what the front side looks like.

In the lower right, the network of resistors makes it possible to compute the voltage at the virtual neutral point. This voltage and the voltages of the three windings are sent to the microcontroller through four of its analog input pins.

Figure 12.16 depicts the rear side of the ESC Shield. This contains the three MOSFET drivers that deliver current to the gates of the MOSFETs.

Differently sized traces carry different amounts of power. That is, traces carrying greater power are wider than those carrying less power. This is why the traces carrying power to the motor are larger than the traces carrying analog signals.

Figure 12.15
ESC Shield board
design (front)

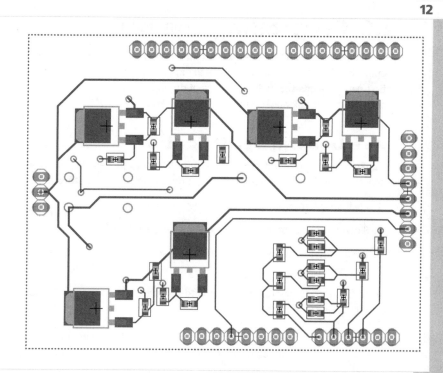

Figure 12.16
ESC shield board
design (rear)

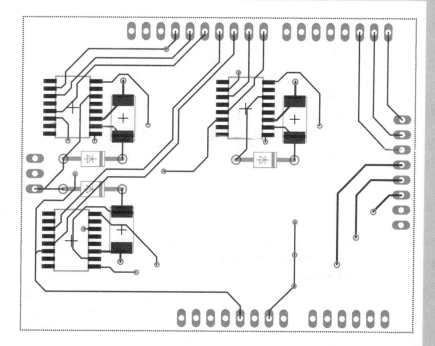

12.6 Controlling the BLDC

At this point, you should have a solid understanding of the ESC Shield circuit and its manner of operation. The goal of this section is to explain how it can be used to provide power to a BLDC. The first part presents the overall process of driving a BLDC, and the second part explains how this process can be approximated with an Arduino sketch.

12.6.1 General BLDC Control

Unlike brushed DC motors, BLDCs require a startup process before they can operate normally. For sensorless BLDCs, this general startup process involves four steps:

1. Position the rotor in a known starting orientation.

2. Slowly turn the rotor clockwise or counterclockwise.

3. Accelerate the rotation until the voltage of the motor's virtual neutral point is large enough to measure.

4. Measure the zero-crossing interval to determine when pulses should be delivered to the motor's windings.

For the first step, the controller delivers current to all three windings to set the rotor's initial orientation. To be specific, it delivers current to the high switch of Winding A, the low switch of Winding B, and the low switch of Winding C. Figure 12.17 shows what this looks like.

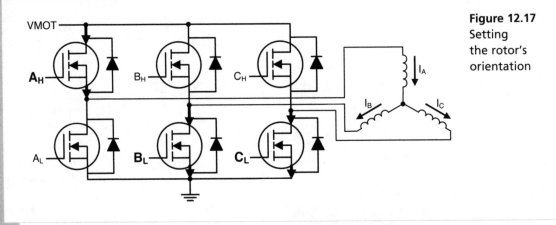

Figure 12.17
Setting
the rotor's
orientation

If the controller provides too much current, the rotor will move rapidly and oscillate around this initial position. Therefore, the controller starts with a low duty cycle and gradually increases it until the rotor's orientation is set.

The next step is to get the rotor moving quickly enough so that its back-EMF will be high enough to allow the controller to perform the zero-crossing measurement discussed earlier. To accomplish this, the controller delivers current to each of the three windings in a staggered sequence. Chapter 3 discussed this sequence, and Figure 12.18 gives an idea of what it looks like.

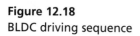

Figure 12.18
BLDC driving sequence

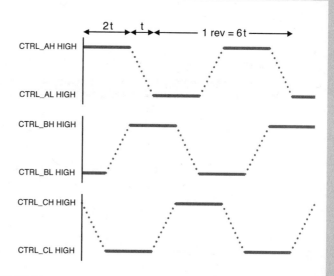

In this diagram, each switch closes for a time equal to 2t, and the winding is left floating for a time equal to t. A complete revolution corresponds to an interval of 6t, so if the desired speed is ω_{goal} (in RPM), the relationship between ω_{goal} and t can be computed as follows:

$$\frac{1\,\text{rev}}{6t\,\text{sec}} = \left(\frac{\omega_{goal}\ \text{rev}}{1\,\text{min}}\right)\left(\frac{1\,\text{min}}{60\,\text{sec}}\right)$$

$$t = \frac{10}{\omega_{goal}}$$

For example, if the goal is to rotate the motor at 400 RPM, the final value of t should be 0.025 sec. However, before the controller sets the final value of t, it needs to monitor the motor's rotational speed and gradually decrease or increase t as needed.

As discussed earlier, the controller determines the motor's speed by measuring the voltage at the motor's virtual neutral point. By comparing this voltage to the voltage in each winding, the controller can measure the zero-crossing intervals to determine how quickly the BLDC is rotating.

12.6.2 Interfacing the BLDC Through Arduino

The 8-bit microcontroller in the Arduino Mega isn't particularly powerful, but it's capable of approximating the BLDC control method discussed in this chapter. Chapter 9 presented the basics of programming the Arduino Mega. Pins 2–13 can be accessed as PWM pins, and six of these pins are needed to provide control pulses to the ESC Shield.

In addition, Pins A0 through A3 make it possible for the Mega to check the zero-crossing of the motor's back-EMF. Table 12.4 lists the Arduino numbers for these pins and the corresponding names of the ESC Shield signals.

Table 12.4   Signal Pins of the ESC Shield

Arduino ID	ESC Shield Signal	Description
13	CTRL_AH	High control signal for Winding A
12	CTRL_AL	Low control signal for Winding A
11	CTRL_BH	High control signal for Winding B
10	CTRL_BL	Low control signal for Winding B
9	CTRL_CH	High control signal for Winding C
8	CTRL_CL	Low control signal for Winding C
A0	ZC_A	Check zero-crossing for Winding A
A1	ZC_B	Check zero-crossing for Winding B
A2	ZC_C	Check zero-crossing for Winding C
A3	ZC_P	Voltage of the virtual neutral point

In the first step of BLDC control, the goal is to bring the rotor to a known orientation. To accomplish this, the controller gradually provides power to the high switch of Winding A (CTRL_AH), the low switch of Winding B (CTRL_BL), and the low switch of Winding C (CTRL_CL). In code, this can be done by calling `analogWrite` and increasing the duty cycle over time.

Once the rotor is in position, the microcontroller starts the motor's operation by delivering power to its windings in sequence. For Winding A, this means providing power to the high switch (CTRL_AH) with the low switch off, then to the low switch (CTLR_AL) with the high switch off, and then to neither switch. The controller delivers a similar sequence to all three windings in a staggered fashion.

As the motor turns, the controller monitors the virtual neutral point. By comparing this value to the voltage of each winding, the controller determines when the winding's back-EMF crosses zero. By measuring the interval between zero-crossings, the controller can determine when to deliver pulses to the motor.

The sketch presented in Listing 12.1 shows how the BLDC control procedure can be approximated on an Arduino board. Note that the ESC Shield must be connected to power, a BLDC, and the headers of a compatible Arduino board.

Listing 12.1 Ch12/bldc.ino—BLDC Control

```
/* This sketch controls a BLDC by applying voltage to the
   six switches on the ESC Shield discussed in Chapter 12 */

// Assign names to the pins
int i, t, va, vp;
int old_time, zc_interval;
int ctrl_ah = 13;
int ctrl_al = 12;
int ctrl_bh = 11;
int ctrl_bl = 10;
int ctrl_ch = 9;
int ctrl_cl = 8;
int zc_a = 0;
int zc_b = 1;
int zc_c = 2;
int zc_p = 3;

int time_goal = 50;

void setup() {

  // Bring rotor to a known initial position
  for (i=0; i<255; i+=5){
    analogWrite(ctrl_ah, i);
    analogWrite(ctrl_bl, i);
    analogWrite(ctrl_cl, i);
    delay(60);
  }

  // Set initial timing value
  old_time = millis();

  // Start turning the rotor slowly
  for (t=500; t>200; t-=50) {
    rotate(t);
  }
  t = 200;
}

void loop() {

  // Check for zero-crossing
  vp = analogRead(zc_p);
  va = analogRead(zc_a);
  if((vp - va < 10) || (va - vp < 10)) {
    zc_interval = millis() - old_time;
    old_time = millis();
```

```
    if(zc_interval - time_goal > 50) {
      t -= 25;
    }
    else if(time_goal - zc_interval > 50) {
      t += 25;
    }
  }

  // Rotate the BLDC
  rotate(t);
}

// Rotate the motor at the given value of t
void rotate(int t) {
  digitalWrite(ctrl_ah, HIGH);
  digitalWrite(ctrl_al, LOW);
  digitalWrite(ctrl_bh, LOW);
  digitalWrite(ctrl_bl, HIGH);
  digitalWrite(ctrl_ch, LOW);
  digitalWrite(ctrl_cl, LOW);
  delay(t);

  digitalWrite(ctrl_ah, HIGH);
  digitalWrite(ctrl_al, LOW);
  digitalWrite(ctrl_bh, LOW);
  digitalWrite(ctrl_bl, LOW);
  digitalWrite(ctrl_ch, LOW);
  digitalWrite(ctrl_cl, HIGH);
  delay(t);

  digitalWrite(ctrl_ah, LOW);
  digitalWrite(ctrl_al, LOW);
  digitalWrite(ctrl_bh, HIGH);
  digitalWrite(ctrl_bl, LOW);
  digitalWrite(ctrl_ch, LOW);
  digitalWrite(ctrl_cl, HIGH);
  delay(t);

  digitalWrite(ctrl_ah, LOW);
  digitalWrite(ctrl_al, HIGH);
  digitalWrite(ctrl_bh, HIGH);
  digitalWrite(ctrl_bl, LOW);
  digitalWrite(ctrl_ch, LOW);
  digitalWrite(ctrl_cl, LOW);
  delay(t);

  digitalWrite(ctrl_ah, LOW);
  digitalWrite(ctrl_al, HIGH);
  digitalWrite(ctrl_bh, LOW);
```

```
digitalWrite(ctrl_bl, LOW);
digitalWrite(ctrl_ch, HIGH);
digitalWrite(ctrl_cl, LOW);
delay(t);

digitalWrite(ctrl_ah, LOW);
digitalWrite(ctrl_al, LOW);
digitalWrite(ctrl_bh, LOW);
digitalWrite(ctrl_bl, HIGH);
digitalWrite(ctrl_ch, HIGH);
digitalWrite(ctrl_cl, LOW);
delay(t);
}
```

The `setup` function performs the first two steps of the BLDC control process. That is, it sets the orientation of the rotor by gradually increasing the duty cycle of the PWM signal delivered to CTRL_AH, CTRL_BL, and CTRL_CL. Then it starts the rotor turning by calling the `rotate` function, which delivers power to the windings in the manner depicted in Figure 12.18.

The `loop` function starts by measuring the voltage at the virtual neutral point, `vp`, and the voltage of Winding A, `va`. If these values are sufficiently close to one another, a zero-crossing is detected. If the motor is rotating slower than the desired speed, the value of the time delay (`t`) is reduced. If the motor is rotating faster than the desired speed, the value of the time delay is increased.

This sketch has one significant flaw. The zero-crossing detection needs to be performed in a separate execution thread than the delivery of power to the windings. Unfortunately, Arduino programming doesn't support threads. Because of this flaw, the ESC Shield is limited in its ability to control BLDCs. Nevertheless, the design process discussed in this chapter can be used to develop general-purpose ESC circuits.

12.7 Summary

When I first learned the difference between brushed motors and BLDCs, I wondered why so many systems still rely on brushed motors. I hope this chapter clarifies why brushed motors remain popular. BLDCs are very difficult to control, requiring significant resources to deliver power to the multiple phases.

A large part of this chapter has focused on the switching circuit that turns motor power on and off. In the ESC Shield, MOSFETS are employed to serve as switches. However, to fully turn on a MOSFET, the gate voltage must be raised higher than the drain voltage and the gate's charge must be moved quickly. To provide high voltage and high current, the ESC relies on MOSFET drivers.

To control a BLDC effectively, the control circuit needs to synchronize its power delivery with the rotor's orientation. Most motors don't have sensors, so circuits must gauge the rotor's orientation by measuring the back-EMF of the BLDC's windings. This chapter explained how measuring the zero-crossing of the back-EMF makes it possible to control the motor effectively.

The final part of this chapter presented the circuit design for the ESC Shield. The schematic design illustrates which components are present in the design and how they're connected. The board design positions the components' positions on the actual board area.

DESIGNING A QUADCOPTER

Within the maker community, one of the most popular applications of electric motors involves spinning the propellers of a quadcopter. Quadcopters and other remotely piloted aircraft have received a great deal of attention in recent years, and corporations are eager to use them to deliver goods and provide ground imagery. At the same time, skeptics worry that their usage threatens personal privacy.

No matter which side of the controversy you take, there can be no question that building a quadcopter is an exciting challenge. In addition to electrical engineering, a designer must be familiar with concepts from mechanical engineering and aeronautical engineering.

I'm not an expert, but I have successfully built and flown a custom quadcopter. This chapter presents my decision-making process and the steps I took to construct the vehicle. Throughout the design process, my primary priorities were simplicity and reliability. The quadcopter presented in this chapter won't break any records, but it flies reliably and doesn't cost nearly as much as a prebuilt professional model.

This chapter discusses each of the different components that make up a quadcopter, starting with the frame. Next, we'll look at selecting the propellers and motors. To assist in the selecting process, I'll derive an expression for a propeller's upward force in terms of its diameter, pitch, and rotation speed.

A significant portion of this chapter is devoted to the electronics needed to power and control a quadcopter. Four subsystems are involved: the receiver, the flight controller, the electronic speed control (ESC), and the battery. Toward the end of the chapter, I'll explain each of these subsystems and show how they're connected together. I'll also do my best to justify the decisions I made in choosing the parts for the vehicle discussed in this chapter.

13.1 Frame

A quadcopter's frame holds the vehicle's electronics, supports the motors and propellers, and sets the overall shape of the system. It should have a landing gear or other mechanism to allow it to land without damaging the system. For these reasons, selecting the frame is one of the most important decisions to make.

The first issue to consider is the material. From what I've seen, quadcopter frames are constructed of one of four materials:

- **Aluminum**—Vibrates, bends on impact, expensive

- **Carbon fiber**—Light, rigid, absorbs vibration, expensive

- **Plastic**—Heavy (depending on the plastic), absorbs vibration, inexpensive

- **Wood**—Absorbs vibration, can break and warp, inexpensive

Carbon fiber is the usual choice for professional and high-performance quadcopters. It provides a great deal of stiffness at low weight, but in some cases, the material may block radio signals.

The chief disadvantage of carbon fiber is the price: Carbon fiber frames generally cost between thousands and tens of thousands of dollars. Therefore, for inexperienced makers and makers on a budget, I recommend plastic frames. They're not as stiff or as light as carbon fiber frames, but they're durable, and can be easily replaced if broken.

For these reasons, I've chosen the Flip Sport frame from hoverthings.com to serve as the frame for the quadcopter discussed in this chapter. This was specifically designed for durability, with arms made of a reinforced plastic called fiberglass. In my usage, it has survived a handful of crashes without suffering any noticeable injury. Figure 13.1 shows what the Flip Sport frame looks like.

Figure 13.1
The Flip Sport quadcopter frame

This frame is easy to assemble and is capable of withstanding impact. Further characteristics are given as follows:

- The frame's mass is 200 grams, or 0.44 pounds.

- The length from one motor shaft to the opposite motor shaft is 385 mm (15.158 inches).

- The price of the frame at http://www.getfpv.com is $89.99.

The frame's shape determines the maximum diameter of the quadcopter's propellers. The frame's designers recommend propellers with diameters of 8, 9, or 10 inches. The following section explores the topic of propellers in detail.

13.2 Propellers

After choosing the quadcopter's frame, the next step is to select the propellers. There are three important factors to consider:

- Diameter

- Pitch

- Material

For the last item, the three main choices for propeller materials are plastic, nylon, and carbon fiber. Carbon fiber is the most expensive, just as it is for the frame. However, carbon fiber propellers are more reasonably priced, so this discussion focuses exclusively on carbon fiber propellers.

The first two items merit greater attention. With so many sizes and shapes available, it's crucial to understand how a propeller's diameter and pitch affect its motion. Therefore, the first part of this section provides a mathematical derivation of how diameter and pitch relate to the propeller's upward force.

13.2.1 Propeller Dynamics

A propeller's job is to produce force. For helicopters and quadcopters, the upward force is commonly called lift or thrust. As designers, we want to select a propeller that will produce enough thrust to raise the quadcopter while using as little power as possible.

As makers, the only factors we control are diameter, pitch, and rotation speed. Therefore, the fundamental question is, how does a propeller's diameter, pitch, and speed relate to its thrust?

If you search on the Internet, you'll find many generally accepted answers to this question. One common equation relates a propeller's power to its pitch, diameter, and speed with the following relationship:

$$P = 1.31 p d^4 \omega^3$$

In this equation, p is the pitch in feet, d is the diameter in feet, and ω is the rotational speed in thousands of RPM. P is the propeller's power in watts.

I searched for the origin of this equation, and to the best of my knowledge, it was first presented in the *Electric Motor Handbook* by Robert J. Boucher. I tracked down a copy of this book and found the equation presented in Chapter 4. However, the book doesn't provide any explanation for how the equation was obtained.

In my opinion, the best fundamental discussion of propeller dynamics (for non-specialists) can be found at http://www.electricrcaircraftguy.com, a blog managed by an aerospace engineer named Gilbert Staples. The following discussion covers the same subject matter as the blog, but I've reworded and restructured the derivation.

Pushing Air Down

In his famous book *Stick and Rudder*, Wolfgang Langewiesche summarizes aerodynamics very simply: *the wing keeps the airplane up by pushing the air down*. A similar statement can be made for propellers and quadcopters. If the thrust of the air pushed downward by the propellers exceeds the system's weight, the quadcopter rises.

Air that isn't affected by a propeller is said to be in the *freestream* state. In this case, the air simply drifts from place to place, and this velocity is denoted by v_0.

The air pushed downward by a propeller is called *downwash*. The velocity of the downwash is called the *exit velocity*, or v_e. Figure 13.2 gives a basic idea of how v_0 and v_e are related.

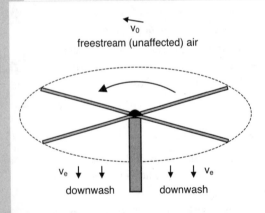

Figure 13.2
Freestream velocity and exit velocity

The velocity of the air near the propeller doesn't change instantaneously from v_0 to v_e. The change in the velocity over time is defined as acceleration, and the downward force exerted by the propeller can be expressed as the product of this acceleration and the mass of the downwash, denoted by m:

$$F = m\left(\frac{v_e - v_0}{t}\right)$$

It's common to rearrange this equation so that force is expressed using the change in the mass over time. This new equation is given as follows:

$$F = \frac{dm}{dt}\left(v_e - v_0\right) = \dot{m}\left(v_e - v_0\right)$$

According to this equation, the velocity of the air above and below the propeller remains constant, but the mass of the downwash changes over time. This change in mass over time is referred to as *mass flow rate*, and is denoted by $\dot{m}$.

Mass Flow Rate

We can't measure the mass flow rate directly, but we can express $\dot{m}$ in terms of quantities that can be measured. The air's mass equals the product of its density (kg/m^3) and its volume (m^3). Denoting density as ρ and volume as V, $m = \rho V$.

Consider the cylinder of air above the propeller. If the propeller's radius is r, this cylinder has a volume equal to $\pi r^2 h$, where h is the height of the cylinder. Figure 13.3 shows what the cylinder looks like.

Figure 13.3
Cylinder of air above a propeller

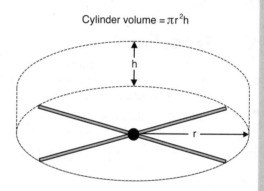

Cylinder volume $= \pi r^2 h$

The mass of air in the cylinder can be expressed with this equation:

$$m = \rho \pi r^2 h$$

Suppose that the mass of air in this cylinder flows through the propeller in time t. This change in mass over time can be expressed in the following way:

$$change = \rho \pi r^2 \left(\frac{h}{t}\right)$$

As h and t approach zero, the fraction becomes the derivative of the cylinder's height with respect to time. This derivative equals the air's exit velocity, v_e. The change in mass equals the mass flow rate, which can now be expressed in the following way:

$$\dot{m} = \rho \pi r^2 v_e$$

Replacing this in the preceding expression for force leads to the following equation:

$$F = \dot{m}(v_e - v_0) = \rho\pi r^2 v_e(v_e - v_0) = \rho\pi r^2\left(v_e^2 - v_e v_0\right)$$

A propeller's length is given in terms of its diameter, d, which is frequently given in inches. To convert this into metric, d must be multiplied by 0.0254 meters/inch. Replacing this in the force equation gives this result:

$$F = \frac{\rho\pi(0.0254d)^2}{4}\left(v_e^2 - v_e v_0\right) = \left(5.067 \cdot 10^{-4}\right)\rho d^2\left(v_e^2 - v_e v_0\right)$$

According to the International Standard Atmosphere (ISA) model, the density of air at sea level at 15°C is 1.225 kg/m3. Replacing this value for ρ produces the following:

$$F = \left(6.207 \cdot 10^{-4}\right)d^2\left(v_e^2 - v_e v_0\right)$$

This assumes that the diameter is given in inches and that the velocities are given in meters per second.

Propeller Pitch

If you take two random screws and start screwing them into a block of wood, one will penetrate further into the wood than the other. The depth of penetration for a single turn is called the screw's pitch. If a screw has a quarter-inch pitch, one turn of the screw will penetrate a quarter of an inch.

A propeller's pitch works in the same way. In theory, a propeller's pitch tells you how far the propeller will move with each revolution. Note that the motion is perpendicular to the propeller's rotation.

Denoting the distance traveled as x, the equation for pitch can be expressed as follows:

$$\text{pitch} = \frac{x}{\text{rotation}}$$

To relate pitch to the force equation, we set the exit velocity equal to the change in x over time. This allows us to solve for v_e in terms of the pitch:

$$v_e = \frac{dx}{dt} = \frac{d(\text{pitch} \cdot \text{rotation})}{dt} = \text{pitch} \cdot \frac{d(\text{rotation})}{dt} = \text{pitch} \cdot \omega$$

Here, ω is the propeller's angular velocity. This is given in RPM, and pitch is usually given in inches. To obtain v_e in metric units (m/s), we need to add conversion factors. Denoting pitch as p, we arrive at the following equation:

$$v_e = p \cdot \left(\frac{0.0254\text{m}}{\text{in}}\right) \cdot \omega \cdot \left(\frac{1\text{min}}{60\text{sec}}\right) = \left(4.233 \cdot 10^{-4}\right)p\omega$$

Replacing this in the earlier force equation gives the following result:

$$F = \left(2.628 \cdot 10^{-7}\right)d^2\left[\left(4.233 \cdot 10^{-4}\right)\left(p\omega\right)^2 - p\omega v_0\right]$$

Gilbert Staples points out that this equation does *not* accurately illustrate the propeller's downward force. In fact, he states that it "doesn't even show the appropriate trends for how thrust changes with varying diameter and pitch."

The problem is that pitch doesn't accurately represent how far the propeller travels per revolution. In other words, v_e is not adequately approximated by dx/dt. To bring the equation in line with experimental data, Gilbert Staples multiplies the force expression by a term involving the inverse of the propeller's pitch ratio (p/d). This produces the following result:

$$F = \left(2.628 \cdot 10^{-7}\right)d^2\left[\left(4.233 \cdot 10^{-4}\right)\left(p\omega\right)^2 - p\omega v_0\right]\left(\frac{d}{3.29546p}\right)^{1.5}$$

The precise expression for the force isn't particularly important, but the form of this equation allows us to draw two important conclusions:

- Of the diameter, pitch, and speed, the diameter has the greatest effect on the resulting force. A small increase in diameter produces a significant increase in force.

- The effect of the propeller's pitch is relatively minor as compared to the diameter and speed.

These conclusions assist in the process of selecting a propeller. The next discussion presents the selection process I employed for the quadcopter.

13.2.2 Selecting the Propeller

As mentioned earlier, the distance between opposite motor shafts on the Flip Sport frame is 385 mm. Therefore, the distance between adjacent motor shafts is 272.24 mm, which equals about 10.72 inches. Figure 13.4 illustrates how these dimensions relate to the quadcopter's frame.

To keep the propeller blades from coming into contact, each propeller must have a diameter less than 10.72 inches. For this reason, this system's propellers have 10-inch diameters.

Surveying the market for propellers with 10-inch diameters, it looks as though the large pitch values lie between 4.5 inches and 4.7 inches. For this reason, I've selected the 10×4.5 (10-inch diameter, 4.5-inch pitch) carbon fiber propeller from Tarot RC Helicopters. They're constructed from carbon fiber, and Figure 13.5 shows what they look like.

Helicopter propellers are specifically marked as CW (clockwise) or CCW (counterclockwise) according to their intended spin direction. The 10×4.5 propellers from Tarot RC Helicopters are sold in pairs, and each pair has one CW propeller and one CCW propeller. When connecting propellers to the quadcopter, make sure that similarly oriented propellers are positioned opposite from one another.

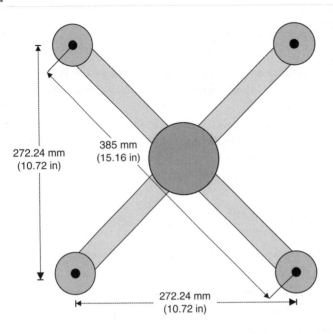

Figure 13.4
Determining maximum propeller size

Figure 13.5
10×4.5 carbon fiber propellers from Tarot RC Helicopters

13.3 Motors

For a normal-sized quadcopter, 10×4.5 propellers are fairly large. This means it takes a significant amount of torque to turn them with enough speed to keep the system aloft. As discussed in Chapter 3, "DC Motors," motors with high Kv values spin quickly but don't provide a lot of torque. Therefore, quadcopters with large propellers should have motors with lower values of Kv.

All quadcopters are powered by battery, so their motors must be DC motors: brushed or brushless. Brushed motors are simpler to control but less efficient and less reliable. Therefore, the design in this chapter makes use of brushless DC motors (BLDC) exclusively. Chapter 3 explains the distinction between brushed and brushless motors in detail.

Another important point to keep in mind when selecting a motor is the diameter of its shaft. A motor's shaft must fit comfortably inside the propeller and the frame. For the Tarot propellers, the diameter of the aperture is 5 mm, so the motor's shaft must have a diameter less than (but not too much less than) 5 mm.

These constraints—BLDC, low Kv, and shaft diameter less than 5 mm—limit the number of suitable motors. Based on these criteria, I've selected the MN3110 KV470 motor from T-Motor, whose website is www.rctigermotor.com. Figure 13.6 shows what these BLDCs look like.

Figure 13.6
The MN3110
K470 BLDC from
T-Motor

The motor's Kv is 470, which is lower than that of most BLDCs on the market. In addition, the diameter of the shaft is 4 mm, which makes it small enough to fit inside one of the propellers from Tarot RC Helicopters. Table 13.1 lists the full set of characteristics for the MN3110 KV470 motors.

Table 13.1 Specifications of the MN3110 KV470 BLDC from T-Motor

Characteristic	Value
Kv	470
Shaft diameter	4 mm
Weight	80 g
Idle current at 10 V	0.3 A

Characteristic	Value
Maximum current	15 A
Configuration	12N14P
Internal resistance	135 mΩ

The motor's electrical characteristics determine what kind of batteries should be employed. This is discussed as part of the next section, which presents how I selected the electronic components of the quadcopter.

13.4 Electronics

Now that the quadcopter's mechanical components have been selected, it's time to look at the electronics. At minimum, a quadcopter's circuitry can be divided into four parts:

- **Transmitter/receiver**—Sends control signals from the user to the quadcopter

- **Flight controller**—Delivers pulses to the ESCs

- **Electronic speed control (ESC)**—Provides power to the motors

- **Battery**—Powers the quadcopter's electronic components

Figure 13.7 shows how control data is transferred among the first three elements. The quadcopter's battery isn't shown in the figure, but it provides power to the receiver, flight controller, and ESCs.

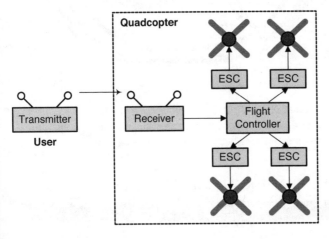

Figure 13.7
Quadcopter control electronics

This section discusses the four elements just listed. In each case, I'll present a set of selection criteria and explain the thought process behind choosing the hardware.

13.4.1 Transmitter/Receiver

Quadcopters are controlled by radio-frequency (RF) signals sent by the user's transmitter. The quadcopter reads the signals using an element called a receiver. In general, transmitters are complex and expensive whereas receivers are simple and cheap.

When you're working with transmitters and receivers, it's important to be familiar with the concept of channels. In RF communication, a channel represents an independent stream of data. When controlling an RC aircraft, each channel manipulates a different actuator. For example, one channel might control the ailerons while another controls the flaps.

For quadcopters, a minimum of four channels are needed. These channels control the vehicle's elevation, roll, pitch, and yaw. In addition, many quadcopters accept additional inputs that control different aspects of the quadcopter's operation. For this reason, most modern transmitters and receivers usually support at least six channels for communication.

Transmitter

When you're selecting a transmitter, an important concern is the mode. A transmitter's mode determines which channels are affected by the left and right stick. The most common mode is Mode 2, in which the left stick controls the rudder and throttle of a traditional RC aircraft. The right stick controls the ailerons and elevator. To make this clear, Figure 13.8 shows which channels correspond to the sticks and switches on a Mode 2 DX6i transmitter from Spektrum.

Other features related to RC transmitters are listed as follows:

- **Model memory**—The transmitter stores settings for different vehicles.

- **Trim**—Enables fine tuning of the vehicle's operation.

- **Programmability**—Enables the transmitter to be configured through a PC connection.

- **Mixing**—Allows channels to be combined so that a pair of control surfaces can be controlled at the same time.

- **LCD display**—Provides information about the transmitter/receiver pairing.

For quadcopters, these features are nice to have but they're not necessary. Therefore, most transmitters for RC aircraft are suitable for quadcopter control.

For this design, I chose the DX6i transmitter from Spektrum. It's more expensive than many of the other receivers, but it provides a wealth of useful features. The main reason I chose the DX6i was availability—I couldn't find any other suitable receivers in stock anywhere.

Receiver

After the transmitter has been selected, the next step is to find a compatible receiver. Compatibility is primarily determined by the transmitter's modulation. Modulation refers to the manner in which the control data is converted to RF signals. Table 13.2 lists three common modulation methods.

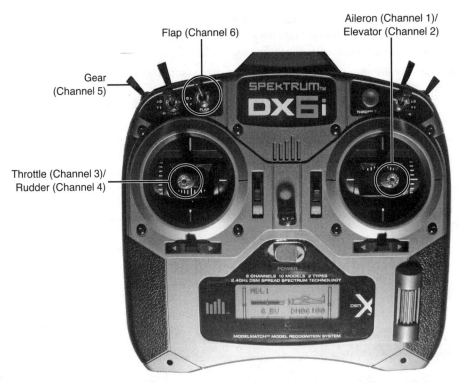

Flap (Channel 6)

Aileron (Channel 1)/
Elevator (Channel 2)

Gear
(Channel 5)

Throttle (Channel 3)/
Rudder (Channel 4)

Figure 13.8
The DX6i
transmitter
from
Spektrum
(Mode 2)

Table 13.2 Common Modulation Methods for Quadcopter Transmitters

Modulation Method	Description
DSM2 (Direct Spectrum Modulation, 2nd generation)	Uses a globally unique identifier (GUID) to bind the receiver to the transmitter
DSMX (Direct Spectrum Modulation X)	Similar to DSM2, but hops between 23 frequencies in the 2.4 GHz band
FAAST (Futaba Advanced Spread Spectrum Technology)	Uses a globally unique identifier (GUID) to identify one of 36 frequency hopping sequences

These methods prevent interference by employing spread spectrum communication methods. In addition, each operates in the 2.4 GHz band, which limits the effective range to the line of sight.

The DX6i transmitter uses DSMX modulation for its control signal, so a compatible receiver must support DSMX. For this reason, I chose the AR610 receiver from Spektrum. Figure 13.9 shows what it looks like.

As shown, the receiver's pins are arranged in a grid containing three rows. With the exception of the BND/DAT column, the pins in the first row provide control signals to the flight controller. The pins in the second and third rows accept positive and negative voltage, respectively.

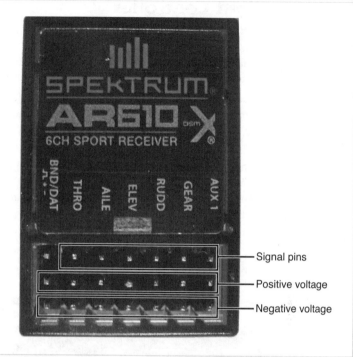

Figure 13.9
The AR610 six-channel receiver

Signal pins

Positive voltage

Negative voltage

Before a receiver can operate normally, it must be bound to a transmitter. That is, the receiver and transmitter must share their IDs so that they recognize one another and can communicate. The binding process depends on the specific receiver and transmitter.

For example, binding the AR610 receiver to the DX6i transmitter requires six steps:

1. Insert a plug across the pins in the AR610's BIND/DAT column.

2. With the DX6i off, move its throttle (the left stick) to the lowest position.

3. Provide power to the AR610 by connecting voltage (3.6 9.6V) across any adjacent positive/negative pins.

4. While holding the Trainer/Bind switch of the DX6i, turn on the transmitter. The LED of the AR610 should start blinking red.

5. Continue holding the Trainer/Bind switch until the AR610's LED becomes constant red.

6. Remove the plug from the AR610's pins.

It's easy to verify that the receiver and transmitter have been bound successfully. When the transmitter is turned on and the receiver has voltage, the receiver's LED should become constant red.

13.4.2 Flight Controller

At minimum, the flight controller's job is to accept input from the receiver and generate control signals for the motors. In addition, many controllers can determine their location through GPS, keep the vehicle level using gyroscopes and accelerometers, and examine the surrounding environment with cameras and atmospheric sensors.

Most flight controllers are proprietary, which means they don't provide any information about their internal design and operation. However, the OpenPilot community, whose main website is www.openpilot.org, has designed a number of open-source flight control circuits that are excellent for quadcopter control.

At present, the two OpenPilot flight controller boards are the Revo (Revolution) and the CC3D (CopterControl 3D). The Revo provides more features, but the OpenPilot store (store.openpilot.org) never seems to have one in stock. Therefore, this discussion focuses on the CC3D, which is illustrated in Figure 13.10.

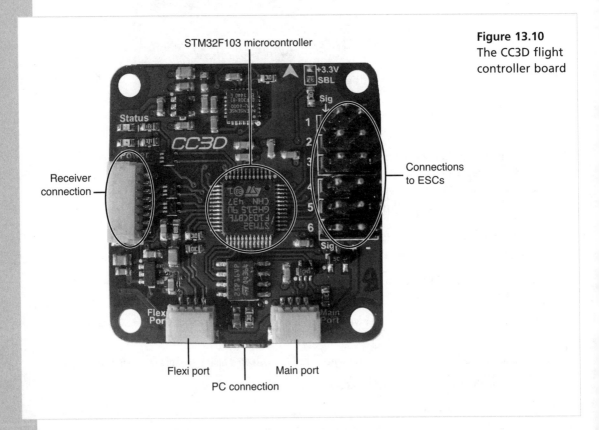

STM32F103 microcontroller

Status

Receiver connection

Connections to ESCs

Flexi port

PC connection

Main port

Figure 13.10
The CC3D flight controller board

Four important features of the CC3D are listed here:

- Receives up to six channels from the receiver

- Processes data and generates pulses with an STM32F103 microcontroller

- Measures the vehicle's motion and orientation with an MPU-6000 six-axis gyroscope/accelerometer

- Stores configuration data with a 16 MB Flash device

If you want to see know a quadcopter really works, it's crucial to understand how the flight controller operates. Figure 13.11 presents a simplified schematic of the CC3D's flight controller.

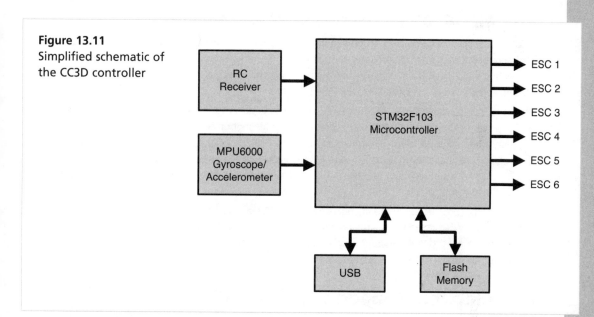

Figure 13.11
Simplified schematic of the CC3D controller

To control the quadcopter, the CC3D makes use of two important devices. The STM32F103 serves as the controller's brain, processing incoming data and delivering control signals to the ESCs. The MPU6000 gyroscope/accelerometer determines the quadcopter's angular orientation and acceleration, and provides the information to the microcontroller.

The microcontroller's program is commonly called firmware. To update the CC3D's firmware and configure its operation, download OpenPilot's Ground Control Station (GCS) software from https://www.openpilot.org/product/openpilot-gcs.

The STM32F103 Microcontroller

Microcontrollers are popular in embedded applications, and Chapter 9, "Motor Control with the Arduino Mega," explained how the Arduino Mega relies on an Atmel microcontroller to control motors. The CC3D processes data with the STM32F103 microcontroller, which serves the same roles as the Arduino Mega's microcontroller, but provides many more capabilities.

The STM32F103 is a 32-bit device that runs at a maximum speed of 72 MHz. Like the Raspberry Pi discussed in Chapter 10, "Motor Control with the Raspberry Pi," it processes data with a processing core from ARM. This core is called the Cortex-M3, and unlike the Raspberry Pi's processor, it's specifically designed for microcontrollers.

The main advantage of using the Cortex-M3 is that it can perform many operations beyond those available for an 8-bit Atmel MCU. The main drawback is that the Arduino programming language isn't available for the STM32F103. So if you want to write applications, you need to understand the C programming language and the microcontroller's architecture.

The MPU6000 Gyroscope/Accelerometer

Most quadcopter pilots don't control each of the four motors individually. Instead, they tell the flight controller to stay level and only change its angular orientation when it's commanded to fly in a direction. For the CC3D, the flight controller determines its orientation by reading data from the MPU6000 gyroscope/accelerometer from InvenSense.

The MPU6000 contains three MEMS (microelectromechanical systems) gyroscopes that identify the device's rate of rotation around the x, y, and z axes. This motion is provided in degrees per second (dps) and the maximum value can be set to 250, 500, 1000, or 2000 dps. In addition, the MPU6000 has three accelerometers that provide the device's acceleration along the x, y, and z axes. The acceleration is provided in terms of the gravitational constant (g), and the full-scale values can be set to 2g, 4g, 8g, and 16g.

On the CC3D, the STM32F103 reads data from the MPU6000 using the serial peripheral interface, or SPI. As the master, the MCU sends commands on the MOSI (Master Output, Slave Input) line. In response, the MPU6000 provides angular rate and acceleration data on the MISO (Master Input, Slave Output) line.

13.4.3 Electronic Speed Control (ESC)

Chapter 3 explained how an ESC makes it possible to control a three-phase BLDC with pulse width modulation (PWM). A quadcopter requires four ESCs, but rather than purchase four separate controllers, it's more convenient to use a single device called a four-in-one ESC. This simplifies wiring and construction, but adds to the overall cost.

A four-in-one ESC has one pair of wires for power and four separate connections to the quadcopter's BLDCs. In addition, it has a 12-pin connector for receiving control signals from the flight controller. Each row of three pins (signal, 5 V, and ground) provides control of one motor.

In addition to cost, there are at least five other issues to consider when selecting an ESC:

- **Battery eliminator circuit (BEC)**—Provides power to the receiver, eliminating the need for a separate receiver battery. Some ESCs have switching battery eliminator circuits (SBECs), also called universal battery eliminator circuits (UBECs), which switch power on and off.

- **Current**—Different ESCs regularly accept different amounts of current, ranging from 20 A per motor to 40 A per motor. It's vital that the ESC can supply enough current to the motors to keep the quadcopter aloft.

- **Weight**—The lower the weight, the better.

- **Programmability**—Most ESCs have microcontrollers that can be configured with new firmware. If an ESC has a microcontroller from Atmel, it can be configured with firmware from Simon Kirby (SimonK), which can be downloaded freely from https://github.com/sim-/tgy. I haven't used it, but many claim that the firmware improves stability and control by increasing the rate at which pulses are sent to the motors.

- **Wire length**—The wires should be long enough to reach all of the quadcopter's motors.

Another significant concern is heat. Low-quality ESCs overheat with high current and cut off power to the motors. Manufacturers don't discuss this in their specifications, so it's important to look at the reviews of any ESC you intend to buy. A popular method of dissipating heat is to glue an aluminum plate to the ESC.

For the quadcopter discussed in this chapter, 25 A of current will be more than sufficient to turn the propellers. For this reason, I've selected the Skywalker Quattro 25Ax4 from Hobbywing. This accepts 25 A of current with bursts of up to 30 A, and provides power to the receiver through a UBEC. Figure 13.12 shows what it looks like.

Figure 13.12
The Skywalker Quattro 25Ax4 ESC from Hobbywing

In my experience, the Skywalker Quattro has continued functioning through many hours of flight. Many reviewers have had similar experiences, but others have reported that the ESC overheated during flight and caused the quadcopter to crash.

13.4.4 Battery

One advantage of having a four-in-one ESC with a BEC is that power only has to be connected to one system. According to the specifications of the Skywalker Quattro 25Ax4, the battery requirements are "2S-4S (7.4V-14.8V)." As discussed in Chapter 3, the "2S-4S" means that sufficient power can be drawn from two to four Li-Po cells connected in series. Li-Po cells are generally 3.7 V, so the total expected voltage ranges from 7.4 V to 14.8 V.

When you're selecting a Li-Po battery, two factors to consider are its capacity and burst rate (C-value). The capacity identifies the total current the battery is capable of supplying at the specified voltage. It's common to see capacity values given in thousands of milliamp-hours (mAh).

The burst rate identifies the battery's maximum rate of discharge. The maximum amount of current that can be drawn from the battery equals its C-value multiplied by its capacity. For example, if a battery with 2100 mAh capacity has a C-value of 20, it can safely discharge a current of 2100 × 20 = 42000 mA = 42 A.

A battery's capacity determines how long current will be available to turn the motor's shaft. However, the greater the capacity, the greater the weight. Therefore, for quadcopters, the increased weight of a high-capacity battery may actually reduce the flying time despite the total amount of current it can provide.

For the quadcopter discussed in this chapter, a 3S Li-Po battery will provide sufficient voltage (11.1 V). Current is a priority, so a high discharge rate is needed. Venom RC makes a 3S Li-Po battery with a capacity of 5,000 mAh and a discharge rate of 35. This means the maximum current is 175 A, which is far more than the quadcopter will need. Figure 13.13 shows what the Venom RC looks like.

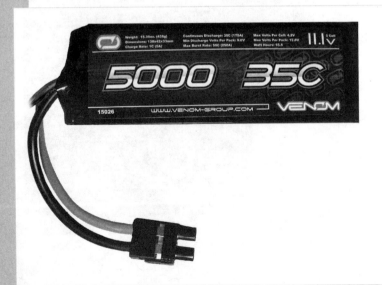

Figure 13.13
The Venom RC 35C 5000 mAh Li-Po battery

This battery provides a handful of different connectors for the power wires, which makes it straight-forward to connect the battery's power to the four-in-one ESC.

13.5 Construction

At this point, most of the quadcopter parts have been selected. But there's one last concern: The frame is too small to comfortably support the electronics, particularly the Venom RC battery. Therefore, I purchased the Tab Center Plate Kit from www.hoverthings.com. This provides two large plates, screws, and standoffs.

After procuring the parts, I constructed the quadcopter by following eight steps:

1. Assemble the Flip Sport frame. Keep in mind that the four arms are connected to the bottom of the center plate.

2. Attach each carbon fiber propeller to a BLDC. Make sure to insert a centering ring inside each propeller before connecting it to the shaft.

3. Attach each BLDC/propeller to an arm of the frame using the four screw holes in the rear of the motor. Make sure the clockwise propellers are positioned opposite one another and the counter-clockwise propellers are positioned opposite one another.

4. The four-in-one ESC fits into the space on the frame's underside between the arms. Attach the ESC to the underside and connect its wires to the four BLDCs. Use zip ties as needed to secure the wires to the frame.

5. The Tab Center Plate Kit contains two cross-shaped plates. Connect one plate to the underside of the frame.

6. Align the battery so that its power wires can reach the power wires of the four-in-one ESC. Sandwich the battery between the two plates of the Tab Center Plate Kit. I used Velcro straps to hold the plates and battery in place.

7. Secure the flight controller and receiver to the top of the frame. Make sure the ESC's control wires can reach the controller.

8. Connect the ESC's control wires to the flight controller and connect the flight controller to the receiver. After these wires are connected, connect the battery's power wires to the ESC.

If the parts have been successfully assembled and connected, the lights on the flight controller should turn on. Figure 13.14 shows what my quadcopter looks like before the ESC wires are connected.

When I started flying the quadcopter, it took some time to get used to the controls. But since then, I've gained a great deal of experience and I look forward to adding more components.

Figure 13.14
The quadcopter

As I said at the beginning of the chapter, I'm not an expert at designing or building quadcopters. Therefore, I recommend that you supplement the material in this chapter with the expert advice from the following sites:

- http://copter.ardupilot.com/wiki/initial-setup/assembly-instructions/hoverthings-flip-sport-quadcopter provides a more in-depth guide to building quadcopters using the Flip Sport frame.

- http://quadcoptersarefun.com/BuildAQuadcopter.html provides many helpful pointers and recommendations for quadcopter components.

- http://www.instructables.com/id/Scratch-build-your-own-quad-copter is an excellent guide for building a quadcopter from scratch.

Lastly, if you run into any issues or difficulties, the forums at www.rcgroups.com and www.hobbyking.com have plenty of knowledgeable people willing to share their expertise.

13.6 Summary

Motor circuits are always fun to operate, but nothing beats the excitement of watching a quadcopter fly. The goal of this chapter has been to explain how to select the parts of a quadcopter and assemble them into a working system.

Given the importance of the quadcopter's propellers, I'm surprised by how little information is available describing how their shape and motion affect the thrust. This chapter has presented a derivation that relates thrust to diameter, pitch, and speed, but this should only be used as an approximate relationship. It's interesting to note that the derivation implies that diameter plays a much larger role than pitch in determining how much thrust a propeller can provide.

Many of the components discussed in this chapter, including the frame, transmitter, and receiver, are matters of taste. However, when you select the motors and propellers, keep in mind that a propeller with a large diameter will require significant torque. To provide the torque, you need a motor with a low value of Kv. It's also important to make sure that the motor's shaft isn't too large for the propeller.

Most flight controllers are proprietary, but the design for the CC3D discussed in this chapter has been released as open source. This circuit board relies on an STM32F103 microcontroller for data processing, and in addition to receiving data from the receiver, it accesses the quadcopter's acceleration and orientation from the MPU6000 gyroscope/accelerometer. The board's design and ground control software can be obtained from www.openpilot.org.

14

ELECTRIC VEHICLES

Of the many applications of electric motors, none is as exciting as their use in automobiles. This chapter focuses on all-electric automobiles, commonly referred to as *electric vehicles*, or *EVs*. EVs provide many advantages over vehicles with combustion engines, including greater reliability, less pollution, less noise, and a decreased reliance on fossil fuels.

A handful of books have already been written on electric vehicles, so this chapter will be limited to providing a broad overview of the topic. I'll start by presenting the topic of EV conversion, which involves replacing a car's combustion engine with an electric motor. In addition to discussing the concepts involved, this section discusses the kinds of motors and batteries commonly used in EV conversion.

The second part of the chapter looks at the state of the art in EV technology. In particular, this section examines three of the most popular EVs on the market as of April 2015: the Tesla Model M, the Nissan Leaf, and the BMW i3. For each vehicle, we'll look at the car's performance, the structure and operation of its motor, and its battery technology.

The last part of the chapter examines four patents from Tesla Motors. In 2014, Elon Musk decided that defending his company's patents was more trouble than it was worth, so he released them into the public domain. Two of the patents discussed in this chapter involve induction motor design, and two others deal with methods of controlling electric motors.

14.1 Electric Vehicle Conversion

Many car owners would like to replace their car's combustion engine with an electric motor. This process, called EV conversion, provides the benefits of owning an electric vehicle without the high cost of buying a new one. In general, EV conversion costs about $10,000, whereas new EVs frequently cost between $30,000 and $80,000.

Few auto mechanics are sufficiently knowledgeable and supportive to convert vehicles from gas to electric. EV conversion shops exist, but they're few and far between. For these reasons, conversion is usually performed by dedicated amateurs.

This section doesn't discuss EV conversion in great depth, but presents a few of the technologies that play a crucial role. To be specific, this section looks at electric motors, controllers, batteries, and transmission.

14.1.1 Motors

When you're converting a car to electric, the motor is one of the most expensive components, commonly costing thousands of dollars. I've only encountered a handful of companies that make motors specifically intended for electric vehicles, and their features and capabilities vary widely.

Choosing an EV's motor is a major decision, so it's important to be familiar with the technology. This discussion starts by explaining how electric motors compare with combustion engines, and then presents the differences between AC and DC motors in electric vehicles. The last part presents different manufacturers of EV motors and their offerings.

Motors, Engines, and Power

When carmakers boast about a new racing engine, the foremost figure of merit is maximum horsepower (hp). As discussed in Chapter 2, "Preliminary Concepts," 1 hp is the power needed to raise 550 pounds in one second. For an average-sized sedan, the engine's maximum horsepower is usually somewhere between 200 and 250 hp.

For electric motors, power is usually expressed in kilowatts (KW), where 1 KW equals 1.341 hp. In most converted EVs, the electric motors put out between 60 and 70 KW, which comes to a range of approximately 80 to 93 hp. As will be shown later in this chapter, the horsepower of a manufactured EV is significantly higher.

This may seem startling, because a combustion-powered 80 hp car will be extremely sluggish, but there are crucial differences between the power of an electric motor and the power of a combustion engine. A combustion engine only reaches its maximum horsepower when it reaches high speed. Its power and torque are dramatically lower at low speed, so it must accelerate to reach its rated horsepower.

As explained in Chapter 2, the input power to an electric motor equals the product of voltage and current. The battery holds the voltage constant, and the current depends on the applied load. This means the power of an electric motor is available immediately after starting. This is why an electric motor can go from 0 to 60 mph in approximately the same time as it takes for a combustion engine with a higher maximum power.

DC Motors and AC Motors

Until recently, AC motors have been considered too expensive and too complex for EV conversion. For these reasons, many sources, including *The Electric Vehicle Conversion Handbook* and *Building an Electric Vehicle*, recommend brushed DC motors. These motors are popular, particularly in forklifts, and they're inexpensive and simple to control.

But as explained in Chapter 3, "DC Motors," the commutator of a brushed DC motor reduces the motor's reliability and efficiency. In addition, it's hard to maintain constant speed as the load changes. This is why vehicles with brushed DC motors have difficulty maintaining speed over hilly terrain.

An important advantage of AC motors involves regenerative braking. A regenerative brake slows a vehicle by using the rotation of the tires to charge the battery. In this case, the motor behaves like a generator, which is the topic of Appendix A, "Electric Generators." Regenerative braking can save a significant amount of power, particularly on long drives, but it's not easily workable for DC motors.

Another advantage of AC motors is that they can maintain torque at varying levels of speed. This means that an AC motor can handle different terrains and acceleration without losing power.

A disadvantage of AC motors involves the complexity of control. It's straightforward to control a brushed DC motor with a battery's DC power, but AC motors require three-phase sinusoidal power. This power can be generated with power inverters, but these systems are more expensive than controllers for brushed DC motor controllers.

Manufacturers

An Internet search for "electric vehicle motors" brings up a handful of companies that make brushed DC motors for EV conversion. These include D&D Motor Systems and Netgain. Netgain's WarP series of motors is particularly popular in EV conversion, and its most recent model, the WarP 11, delivers 34.4 hp of power at 144 V.

For AC motors, two highly respected sources are Siemens and BRUSA. Siemens produces a wide range of AC motors, but from what I've seen, they specialize in selling to EV manufacturers instead of conversion enthusiasts. BRUSA sells its motors online, but the least expensive motor I've encountered was over $11,000.

A third company, called Hi Performance Electric Vehicle Systems (HPEVS), provides AC motors at prices between $3,000 and $6,000. Table 14.1 lists four of its offerings and presents the power, weight, and voltage for each.

Table 14.1 Specifications of AC Electric Motors from HPEVS

Motor	Power	Weight	Voltage
AC-34	65 hp @ 2900 RPM	85 lb/38.5 kg	48/72/96/108 V
AC-35	63 hp @ 2900 RPM	85 lb/38.5 kg	48/72/96/108/144 V
AC-50	71 hp @ 2900 RPM	115 lb/52.2 kg	48/72/96/108/144 V
AC-51	88 hp @ 2900 RPM	115 lb/52.2 kg	96/108/144 V

The main site for HPEVS is http://www.hpevs.com. Looking through its offerings, you may notice some motors with the "X2" designation. This implies that two matched motors can be inserted into an electric vehicle. This dual-motor configuration delivers more horsepower but consumes more current from the battery.

14.1.2 Controllers

For brushed DC motors, control circuits are simple to design and understand. The first concern is power delivery. The controller must be able to deliver enough power to the motor to ensure suitable driving while preventing too much power from damaging the electronics.

The second concern involves controlling the motor's torque and speed. When comparing controllers, you may want to return to Chapter 3 to review the different types of brushed DC motors:

- **Permanent magnet**—Constant magnetic field

- **Series-wound**—Improved torque, worse speed control

- **Shunt-wound**—Good speed control

- **Compound**—Combination of series-wound and shunt-wound

Each controller is best suited for a specific type of motor. Netgain provides controllers for its WarP motors. D&D Motor Systems and Curtis Instruments also sell controllers for brushed DC motors.

To drive an AC motor, the controller needs to convert battery power to three-phase sinusoidal power. For this reason, the terms *AC controller* and *power inverter* are generally synonymous, but in most EVs, AC controllers do more than just deliver AC power. Many inverters can be tuned to optimize the power frequency, and some can be programmed to support different modes of operation.

Many different types of AC controllers are available, but two companies provide controllers specifically intended for electric vehicles. Curtis Instruments sells a number of AC controllers, and many conversion enthusiasts have reported they work well with HPEVS motors. In addition, BRUSA provides controllers that operate with high performance and high efficiency.

14.1.3 Batteries

The type and number of batteries is another major concern in the EV conversion process. From what I've seen, EV batteries come in one of three types:

- **Lead-acid**—Low cost, high weight

- **Lithium-ion**—Low weight, high cost

- **Lithium-iron-phosphate (LFP)**—Low weight, high safety, high cost

The specific energy of lead-acid batteries (33–42 Wh/kg) is significantly lower than the specific energy of lithium-ion batteries (100–265 Wh/kg). This means an EV with lead-acid batteries will be significantly heavier than an EV with the same amount of lithium-ion batteries.

Despite this, many converted vehicles rely on lead-acid batteries because of cost. A 12 V 80 Ah lead-acid battery costs about $300, whereas a 12V 80 Ah lithium-ion battery costs about $1,000.

However, lead-acid batteries become unusable after approximately 600 cycles, whereas lithium-ion batteries can last approximately 1,000 cycles. When selecting EV batteries, it's important to keep these trade-offs in mind.

Another trade-off involves safety. In 2013, three Tesla Model S vehicles caught fire because their lithium-ion batteries were damaged. Manufacturers are taking pains to ensure that these batteries are kept safe from harm, but lithium-ion chemistry is inherently volatile. For this reason, some EV conversion experts strongly recommend lithium-iron-phosphate (LFP) batteries, which are more stable than lithium ion batteries but provide less specific energy.

After the type of battery has been chosen, the next step is to determine how many are needed. This is determined by the requirements of the controller and the motor. For example, if the controller requires 96 V, it will take eight 12 V batteries connected in series to provide the necessary voltage. The current to the controller can be doubled by adding another eight batteries in parallel.

14.1.4 Transmission

An automobile's transmission is the mechanical system that transfers the motor's/engine's power to the wheels. When it comes to electric vehicles, there are two main issues related to the transmission: connecting the electric motor to the transmission and the need for manual transmission over automatic transmission.

The structure of an automobile's transmission changes from vehicle to vehicle. Therefore, to connect an electric motor to the transmission, you need an adapter specifically built for that motor and vehicle. EV conversion kits provide prebuilt adapters, but in many cases, a custom adapter needs to be fabricated. This can be a complex process, and for an in-depth presentation, I recommend *Building an Electric Vehicle* by Ken Watkins.

A vehicle with an automatic transmission will automatically change gears as it moves. They're available in most automobiles with combustion engines, but they're not practical in EVs. This is because automatic transmissions significantly reduce the horsepower provided by the motor. Further, they add a layer of complexity that most EV conversion amateurs aren't willing to deal with. For this reason, nearly all converted EVs have manual transmissions.

14.2 Modern Electric Vehicles

If you want to understand the current state of EV technology, the best place to look is the marketplace. As I write this in April 2015, the most popular EVs are the Tesla Model M, the Nissan Leaf, and the BMW i3. This section provides a technical overview of each car, with particular emphasis on the motor and battery.

14.2.1 Tesla Motors Model S

In 2008, Tesla Motors released the Tesla Roadster, the first EV to be taken seriously by auto connoisseurs. In 2012, it released the Tesla Model S, which improves on the Roadster in many respects. Figure 14.1 shows what the Model S looks like.

Figure 14.1
The Tesla Model S

The Model S, like all EVs from Tesla, relies on an induction motor to provide motion. As discussed in Chapter 6, "AC Motors," an induction motor is an AC motor whose cylindrical rotor has conductive plates mounted on its perimeter. As AC current enters the stator, the magnetic field induces voltage in the rotor. The interaction between the rotor's induced voltage and the stator's field produces motion. As a review, Figure 14.2 shows what the rotor of a squirrel cage induction motor looks like.

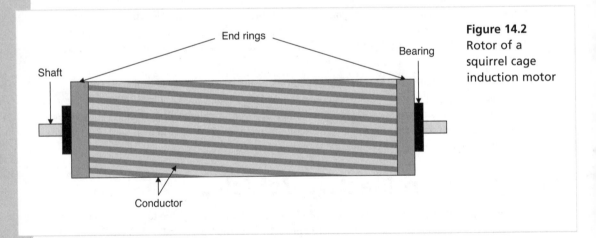

Figure 14.2
Rotor of a squirrel cage induction motor

In 2007, Wally Rippel, a Principal Power Electronics Engineer at Tesla Motors, posted a message on the Tesla Motors blog titled "Induction Versus DC Motors." He discusses the technology underlying induction motors and brushless DC motors (BLDCs), and explains why Tesla Motors prefers induction motors for its vehicles. His message cites two main reasons:

- The rare-earth permanent magnets in a BLDC, such as dysprosium and neodymium, are prohibitively expensive.

- A BLDC's magnets produce a constant magnetic field. In an induction motor, the magnetic field is determined by V/f. This means the electrical losses in an induction motor can be controlled by reducing V/f as needed.

The choice of motors in the Model S has yielded impressive performance results. The base Model S runs at a maximum of 362 hp and the Performance Model S runs at 416 hp. Table 14.2 provides further performance numbers for different variants of the Model S.

Table 14.2 Performance Characteristics of the Tesla Model S

Model	Battery Capacity	Equivalent Miles Per Gallon (City)	Equivalent Miles Per Gallon (Highway)
Model S	60 KWh	94	97
Model S	85 KWh	88	90
Model S AWD	85 KWh	86	94
Model S AWD-85D	85 KWh	95	106
Model S AWD-P85D	85 KWh	89	98

Model S batteries are all based on lithium-ion technology. Currently, the largest battery has a capacity of 85 KWh, which allows the vehicle to travel 265 miles (426 km) before having to recharge. Tesla Motors also provides a 70 KWh battery option, which replaces its old 60 KWh battery option.

14.2.2 Nissan Leaf

In 2010, Nissan Motor Company released the Nissan Leaf, a compact hatchback that runs on battery power. It won the World Car of the Year Award in 2011, and Figure 14.3 shows what it looks like.

Figure 14.3
The Nissan Leaf

Like the Model S, the Leaf relies on an AC motor to provide motion. This means it uses an inverter to convert the battery's DC power to three-phase sinusoidal power.

In contrast with the induction motor of the Model S, the Leaf's motor is a synchronous AC motor. In this case, the rotor generates a magnetic field using permanent magnets. Figure 14.4 illustrates the rotor and stator of a synchronous AC motor.

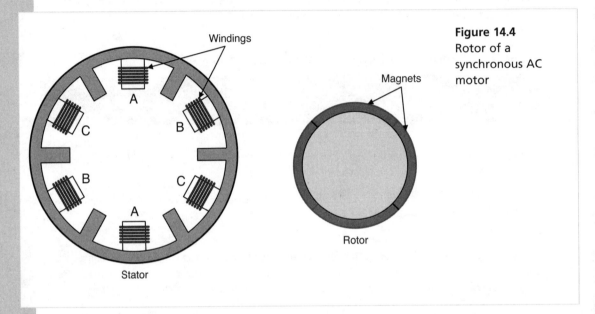

Figure 14.4 Rotor of a synchronous AC motor

Windings

A

C   B

B   C

A

Stator

Magnets

Rotor

In November 2012, Steve Rousseau wrote a short article about the Nissan Leaf's motor for *Popular Mechanics*. The topic involves the rare-earth metals that serve as the rotor's permanent magnets. According to the article, Nissan's scientists have devised a new process for combining the rare-earth metals dysprosium and neodymium, thereby reducing the amount of dysprosium needed by 40%. This gives an idea of how important rare-earth elements are for electric motors in vehicles such as the Leaf.

With its synchronous motor, the Nissan Leaf runs at a peak horsepower of 110 hp. This enables it to accelerate from 0 to 60 mph in 9.9 seconds. For the 2010/2011 model, the top speed has been measured at 93 mph, or 150 kph.

For power storage, Nissan provides a 24 KWh lithium-ion battery. According to the U.S. Environmental Protection Agency, the range of the Nissan Leaf is 73 miles, or 117 km.

Despite having less battery capacity and a lower-power motor, the Nissan Leaf is currently outselling the Model S across the world. The reason is pricing. The manufacturer's suggested retail price (MSRP) of the base 2015 Nissan Leaf is $29,680. The MSRP of the base 2015 Tesla Model S is $69,900.

14.2.3 BMW i3

In 2013, BMW released the i3, the first of its "Project i" line of vehicles. This EV won the World Car Design of the Year award in 2014, and Figure 14.5 shows what it looks like.

Figure 14.5
The BMW i3

The BMW i3 resembles the Nissan Leaf in a number of respects. Both rely on synchronous AC motors to provide mechanical power, and both have relatively small batteries. This small battery size reduces the car's price and weight, but also reduces the distance it can travel between charges.

At minimum, the i3 relies on a 170 hp synchronous AC motor to turn the wheels. This allows the vehicle to accelerate from 0 to 60 mph in under eight seconds. The i3's maximum speed is 93 mph.

In addition to the electric motor, the i3 can have an optional gasoline engine called a range extender, or REx. This is essentially a motorcycle engine, and though it increases the vehicle's speed and range, it requires stops for fuel as well as electricity.

For power, the i3 relies on a 22 KWh lithium-ion battery. The vehicle's range depends on its operating mode, which can be set to one of three options:

- **COMFORT**—Range of 80–100 miles

- **ECO PRO**—Reduces air conditioning, limits maximum speed to 80 mph, and extends range by 12%

- **ECO PRO+**—Completely deactivates air conditioning, limits maximum speed to 55 mph, and extends range by 24%

One important advantage of the BMW i3 relates to the materials used in its construction. The i3's frame is made out of carbon fiber reinforced plastic, or CFRP. This dramatically reduces the vehicle's weight without sacrificing its durability. This explains why the i3's range is longer than you'd expect from a comparable EV with a 22 KWh battery.

14.3 Patents from Tesla Motors

In June 2014, Elon Musk posted an announcement on the Tesla Motors blog titled "All Our Patent Are Belong To You." He expressed his disappointment with patents, stating that they "stifled innovation" and that "receiving a patent really just meant that you bought a lottery ticket to a lawsuit."

To encourage the development of electric vehicles, he decided that all patents awarded to Tesla Motors would be freely available to the public. Or as he put it, "Tesla will not initiate patent lawsuits against anyone who, in good faith, wants to use our technology."

Now that these patents are essentially in the public domain, I selected four that I found particularly interesting:

- **Patent 7,960,928**—Controlling motor operation based on user input and/or vehicle information
- **Patent 8,154,167**—A new design for induction motor lamination
- **Patent 8,453,770**—Control system for dual-motor vehicles
- **Patent 8,572,837**—Method for making efficient rotors in electric motors

Most of the technical content in this book, such as the operation of an AC motor, has been known for nearly a century. However, these patents represent the latest state of the art, and if Tesla Motors considered them worthy of patent protection, they're worth our examination.

14.3.1 Flux Controlled Motor Management

Patent 7,960,928 introduces a method of motor control that acquires data from a variety of inputs. The patent presents five operational modes that determine the motor's behavior:

- **Performance mode**—The controller optimizes the vehicle's torque and speed.
- **Efficiency mode**—The controller optimizes the vehicle's fuel efficiency.
- **Regenerative mode**—The controller optimizes the energy recovered through regenerative braking.
- **Thermal mode**—The controller keeps the vehicle's internal temperature to a minimum.
- **Traction mode**—The controller maximizes the tires' grip on the road.

The patent recognizes that more than one mode may be selected at any given time. Most of the patent's description focuses on how the motor's speed and torque should be controlled in each case. For example, the section on thermal management explains how the motor's heat is determined by the currents in the rotor and stator. Therefore, if thermal mode is selected, the controller needs to keep the current to a minimum.

The vehicle's on-board electronics determine which modes are active at any given time. The following text, taken directly from the patent, gives an idea of how this works:

"...when a throttle (or torque command input) is hard pressed, the vehicle settings can change (e.g. automatically or manually) to the maximum performance mode (e.g., still using the first regeneration mode). Then when highway cruising is detected or selected, the vehicle settings can change back to the maximum efficiency mode (e.g. still using the first regeneration mode). When highway passing is detected or desired, the vehicle settings can change back to maximum efficiency mode...."

Figure 14.6 provides a graphical illustration of the controller's overall process for managing the motor's operation.

Figure 14.6
Motor control decision-making process

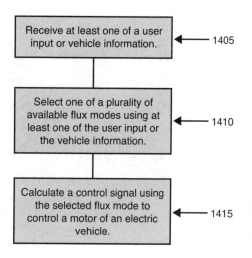

Receive at least one of a user input or vehicle information. ◄— 1405

Select one of a plurality of available flux modes using at least one of the user input or the vehicle information. ◄— 1410

Calculate a control signal using the selected flux mode to control a motor of an electric vehicle. ◄— 1415

Controlling the AC motor in a commercial system is a complex process in the best of cases, but reading this patent gives an idea of how complicated it is to manage the operation of an EV motor.

14.3.2 Induction Motor Lamination Design

Patent 8,154,167 presents an innovative design for AC induction motors. This is essentially a variation of the squirrel-cage rotor discussed in Chapter 6. That is, the cylindrical rotor has conductors that receive induced voltage from the stator, but no permanent magnets.

This patent presents a new design for an induction motor's rotor and stator. Figure 14.7 illustrates the cross-section.

The patent is very specific about the geometry of the features depicted in the figure. For example, the patent's claims include the following:

- The motor should have 60 stator teeth and 74 rotor teeth.

- The length of each rotor/stator tooth should be 4–6 times its width.

- The length of each rotor tooth should be 1–1.2 times the length of each stator tooth.

- The air gap between the rotor and stator should be kept between 0.5 and 0.8 millimeters.

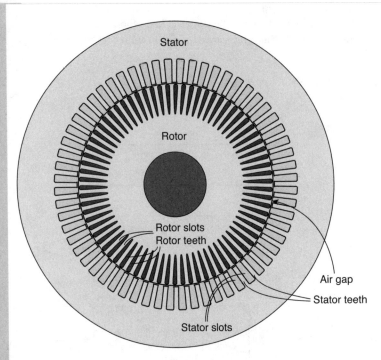

Figure 14.7
Cross-section of the patented induction motor design

The windings in each stator slot are separated into two layers. These windings carry current in three phases, denoted as A, B, and C. Figure 14.8 gives an idea of how the phases are organized in the stator's 60 slots.

The patent explains how the winding coils can be inserted in the stator slots to enable automatic manufacturing:

> "...due to the winding approach provided by this invention, the inter-pole connections are accomplished during the winding and coil insertion process, not after coil insertion as is common in prior art winding patterns ... The elimination of the post-insertion inter-pole connection steps simplifies motor production, thus reducing cost and motor complexity while improving motor reliability and quality."

The patent description ends by discussing the simulated electromagnetic fields surrounding the rotor and stator. The design and placement of the teeth are specifically chosen to optimize the induced current in the rotor's conductors.

Figure 14.8
Two layers of windings in the induction motor design

1st Layer

	Slot																													
	1	2	3	4	5	6	7	8	9	10	11	12	13	14	15	16	17	18	19	20	21	22	23	24	25	26	27	28	29	30
Upper	A1	A1						B4	B4					A1	A1	A2	A2				C1	C1							A2	A2
Lower	A1	A1	A1					B4	B4	B4			A1	A1	A1	A2	A2	A2			C1	C1	C1					A2	A2	A2

	Slot																													
	31	32	33	34	35	36	37	38	39	40	41	42	43	44	45	46	47	48	49	50	51	52	53	54	55	56	57	58	59	60
Upper			C1	C1	C2	C2			B3	B3									C2	C2				B3	B3	B3	B4	B4		
Lower			C1	C1	C1	C2	C2	C2			B3	B3	B3						C2	C2	C2			B3	B3	B3	B4	B4	B4	

- -

2nd Layer (in bold & italics)

	Slot																													
	1	2	3	4	5	6	7	8	9	10	11	12	13	14	15	16	17	18	19	20	21	22	23	24	25	26	27	28	29	30
Upper	A1	A1	*C3*	*C3*	*C3*	*C4*	*C4*	*C4*	B4	B4	*B1*	*B1*	*B1*	A1	A1	A2	A2	*C4*	*C4*	*C4*	C1	C1	*B1*	*B1*	*B1*	*B2*	*B2*	*B2*	A2	A2
Lower	A1	A1	A1	*C3*	*C3*	*C4*	*C4*	B4	B4	B4	*B1*	*B1*	A1	A1	A1	A2	A2	A2	*C4*	*C4*	C1	C1	C1	*B1*	*B1*	*B2*	*B2*	A2	A2	A2

	Slot																													
	31	32	33	34	35	36	37	38	39	40	41	42	43	44	45	46	47	48	49	50	51	52	53	54	55	56	57	58	59	60
Upper	*A3*	*A3*	*A3*	C1	C1	C2	C2	*B2*	*B2*	*B2*	B3	B3	*A3*	*A3*	*A3*	*A4*	*A4*	*A4*	C2	C2	*C3*	*C3*	*C3*	B3	B3	B4	B4	*A4*	*A4*	*A4*
Lower	*A3*	*A3*	C1	C1	C1	C2	C2	C2	*B2*	*B2*	B3	B3	B3	*A3*	*A3*	*A4*	*A4*	C2	C2	C2	*C3*	*C3*	B3	B3	B3	B4	B4	B4	*A4*	*A4*

14.3.3 Dual-Motor Drive and Control System

Most EVs have one motor, but Patent 8,453,770 is concerned with controlling vehicles with two motors. Dual-motor vehicles have a number of advantages over single-motor EVs:

- Superior power optimization and system efficiency.

- Load balancing ensures that motors operate in their optimal temperature range.

- Weight distribution in the vehicle design is simplified.

- The control system can be optimized for different environments and operating modes.

Another advantage of dual-motor systems is that one motor can be used for regular, low-speed usage while the other is reserved for high-speed or high-load conditions. This improves the vehicle's efficiency and lengthens the operational life of the two motors.

A disadvantage of having two motors is that the controller has to perform additional computation to generate control signals. Patent 8,453,770 focuses on the decision-making process employed by the controller, which it refers to as the *torque control unit*. This control unit consists of three parts:

- **Traction control command generation unit**—Computes speed, wheel slip ratios, and slip errors

- **Torque split unit**—Computes optimal torque requests for each motor

- **Traction control unit**—Minimizes slip errors

To perform its function, the torque control unit receives input from multiple sensors on the vehicle, including speed sensors, steering sensors, brake sensors, and gear-selection sensors. Figure 14.9 gives an idea of how the sensors deliver data to the controller, and how the controller interacts with the rest of the vehicle.

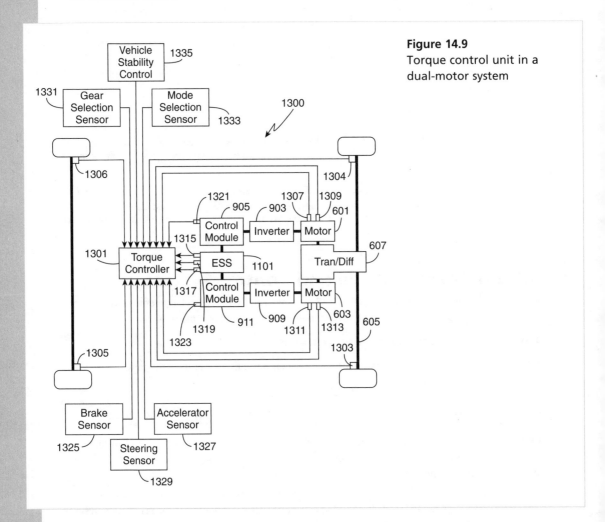

Figure 14.9
Torque control unit in a dual-motor system

The patent's claims go into detail regarding the type of control methods employed by the controller's circuitry. For example, the traction control unit relies on a second stage feedback loop to reduce speed disturbances in both motors. It may also use a feedforward control circuit to manage torque production.

In the figure, ESS stands for energy storage system, which could be a battery or supercapacitor. The ESS has sensors for temperature, voltage, and current, and they provide data to the torque-limiting unit. These torque-limiting units compute the maximum torque that can be exerted by the two motors. The torque control unit uses this information to determine how much torque should be provided by each motor.

14.3.4 Method for Making Efficient Rotors

Whereas the preceding patents have dealt with motor design and control, Patent 8,572,837 is concerned with manufacturing. Specifically, it explains how to efficiently construct the cylindrical rotor of an AC induction motor.

As discussed in Chapter 6, the rotor of an induction motor turns because of currents induced in its conductors by the stator's field. A chief issue with manufacturing this motor involves inserting the conductors into the rotor so that they're equally spaced but only touch one another at the rotor's ends. Another issue involves designing the rotor so that it can be manufactured by an automatic process. Figure 14.10 shows what the proposed process looks like.

Figure 14.10
Method of manufacturing the rotor of an induction motor

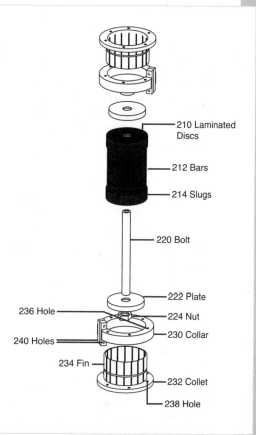

The central portion of the rotor is composed of laminated steel discs with grooves that extend radially outward. Conductive bars, usually made of copper or aluminum, are inserted into the grooves, and each bar runs the length of the cylinder. On either end of the cylinder, the bars come in contact with one another through conductive slugs. The plate and collar on either end are used to compress the discs and clamp the assembly in place. In addition, beryllium bands may be used to compress the slugs.

The central concern of the patent is the rotor's conductive bars. Lowering the bars' resistance increases the motor's efficiency, and because copper is a better conductor than aluminum, the patent focuses on copper bars. To electrically connect the bars with low resistance, the patent states that the slugs should be made from silver-plated copper.

Silver has a low melting point, so when the rotor assembly is heated in the furnace, the silver melts, improving the contact between the bars and the slugs. When the assembly is cooled, the rotor's bars and slugs are fully connected. The plates and collars are then removed and the rotor can be inserted into an induction motor.

14.4 Summary

The importance of electric vehicles can't be overstated. In addition to reducing pollution, they help liberate society from its dependence on fossil fuels. At the time of this writing, EVs are too expensive and insufficiently powerful to threaten traditional vehicles in the marketplace, but the day of their ascendance is rapidly approaching.

For many, the only practical way to get involved in EV technology is to convert an existing vehicle. EV conversion is expensive, but not as expensive as purchasing a new EV. The primary decisions to make involve selecting the motor and the battery. Brushed DC motors are inexpensive and easy to control, but are inefficient, unreliable, and tend to have issues with speed control. AC motors are more expensive and complex, but they allow for regenerative braking, which means the vehicle's motion can recharge the battery during travel.

Regarding battery selection, lead-acid batteries are inexpensive but provide little current capacity for the weight. Lithium-ion batteries provide excellent current capacity, but are expensive and can be volatile if mishandled. Lithium-iron-phosphate batteries don't provide as much current capacity per weight as lithium-ion batteries, but are generally safer and easier to manage.

The second part of this chapter discusses three electric vehicles currently available in the marketplace: the Tesla Motors Model S, the Nissan Leaf, and the BMW i3. The Model S provides more horsepower than the other offerings as well as greater current capacity, but the price tag is 2–3 times as large.

The last part of the chapter presents four patents that Tesla Motors recently released into the public domain. These patents provide a fascinating degree of insight into the cutting edge of electric motor technology. Two of the patents illustrate methods of controlling motors in electric machines. It's informative to see how many factors go into the controller's decision-making process.

All the EVs from Tesla Motors rely on induction motors, so it's no surprise that many of the company's patents relate to induction motors. Of the patents discussed in this chapter, one presents the design of an induction motor with teeth on the rotor and stator. Another presents a method of manufacturing the rotor of an induction motor.

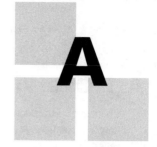

ELECTRIC GENERATORS

The chapters in this book have focused on motor action, which converts electrical power to mechanical power. This appendix takes a look at the reverse process, which converts mechanical power to electrical power. This is called generator action, and machines that take advantage of this action are called electric generators. As will be shown, motors and generators have a lot in common.

Electrical generators serve a vital role in today's society. When catastrophe strikes, the first priority is to return the generators to working condition. In addition to providing power to businesses and residences, generators also make it possible for automobiles to power their electrical components, such as the headlights, power steering, and power locks.

To present this topic, this appendix divides generators into two categories: AC and DC. Both types have similar internal structures, but DC generators have to do extra work to make the output power suitable for DC circuits. Generators of DC power are called dynamos, and the first part of this appendix looks at how they work and how they rectify and smooth the output.

AC generators are much more common than DC generators, and most of this appendix is devoted to this topic. In particular, the category of AC generators will be further subdivided into those that rely on permanent magnets and those that don't. The generators in the first category are called magnetos and are used in applications requiring simplicity and high reliability. The generators in the second category are called self-excited generators, and are used in modern vehicles and power production facilities.

A.1 Overview

Chapter 1, "Introduction to Electric Motors," presented the basics of electric motors and showed how a current-carrying conductor in the presence of an electric field receives a force. This force, called the Lorentz force, is responsible for the motion of electric motors.

The reverse phenomenon can also take place. That is, when a conductor without current moves inside a magnetic field, it receives an induced voltage that produces a current. As discussed in Chapter 6, "AC Motors," this is how induction motors work. However, in a generator, the motion is the input and the induced voltage is the output.

In a generator, the motion may involve the conductor or the field-producing element. As with motors, the moving element is called the rotor and the stationary element is called the stator. Further, the element receiving the induced voltage is called the armature and the element that generates the magnetic field is the field magnet or field winding.

Figure A.1 gives an idea of how a basic generator works. In this case, the conductive armature rotates inside the poles of a magnet. As it rotates, the induced voltage causes current to flow through the wire.

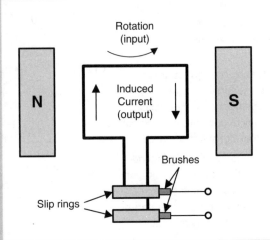

Figure A.1
A simple electric generator

The input mechanical power turns the conductor inside the magnet. The rotation induces a voltage inside the conductor that produces current. The current reaches the external circuit through slip rings, which are connected to contacts called brushes.

The input power equals the product of the rotor's torque, τ, and rotational speed, ω. The output power equals the product of the induced voltage, V, and the resulting current, I. Therefore, the efficiency of a generator can be expressed with the following equation:

$$\eta = \frac{P_{output}}{P_{input}} = \frac{P_{electrical}}{P_{mechanical}} = \frac{VI}{\tau\omega}$$

The generator depicted in this figure may produce enough electricity to light a small bulb, but not much more. Generators need to be large to produce useful amounts of electricity. For example, if you convert a stationary bicycle into an electric generator and start pedaling, you'll probably only generate between 80 W and 100 W, which is barely enough power to run a small television.

Just as motors can run on DC or AC electrical power, electrical generators can produce DC or AC electric power. The rest of this chapter presents the generators of both types.

A.2 DC Generators

Through much of the nineteenth century, power was transferred as direct current instead of the alternating current we use today. This made it necessary to build generators of DC electricity, which were originally called dynamo-electric machines or simply dynamos.

 note

Dynamos aren't the only type of DC electrical generators. Homopolar generators and magnetohydrodynamic (MHD) generators also produce DC power, but these topics lie beyond the scope of this discussion.

A dynamo's operation is essentially similar to that of the generator depicted in Figure A.1. In fact, a dynamo's structure looks a lot like that of the brushed DC motor presented in Chapter 3, "DC Motors." Both have commutators, permanent magnets, and brushes that connect wires to the rotating conductor. The main difference between the two is that the nature of the input/output power is reversed: Dynamos accept mechanical power as input and produce electrical power as output.

As the rotor completes a rotation, the current produced by the dynamo grows and diminishes. The commutator switches the direction of the current every half-turn, thereby ensuring that the current always flows in the same direction. Figure A.2 shows what the resulting current looks like.

Figure A.2
Rectified power produced by a dynamo

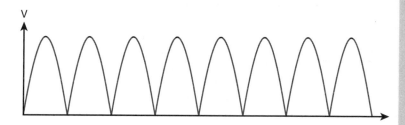

Before this output can be used to power a DC system, it must be brought to a more constant level through a process called smoothing. This can be accomplished by placing a capacitor in parallel with the load. This is called a smoothing capacitor or a reservoir capacitor, and the higher the capacitance, the smoother the output. Figure A.3 shows what the circuit looks like.

Figure A.4 shows what the resulting power looks like. The variation in the output voltage is referred to as ripple.

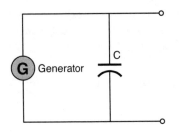

Figure A.3
Simple smoothing circuit

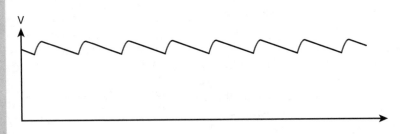

Figure A.4
Smoothed voltage with ripple

If a smoothing capacitor won't be sufficient, the ripple can be reduced further with a filter made up of two capacitors and an inductor. This is called a pi filter, and Figure A.5 shows what it looks like.

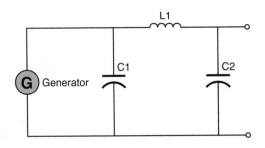

Figure A.5
Pi filter for improved smoothing

In this circuit, the first capacitor (C1) looks like a short circuit for AC current and prevents a large part of the input's AC component from reaching the output. The impedance presented by the inductor (L1) further diminishes the AC component passing through it. The second capacitor (C2) carries AC power away from the output, which should now be sufficiently smooth for most applications requiring DC power.

A.3 AC Generators

Generating DC power was a major concern in the nineteenth century, but in time, all large-scale electrical machinery came to rely on AC power. The reason has to do with transformers. Transformers make it possible to step up (increase voltage, reduce current) or step down (reduce voltage, increase current) AC power. Low current means low transmission losses, so modern power systems apply step-up transformers before transmission. When the destination is reached, a step-down transformer increases the current to a usable level.

Because AC power can be transmitted with minimal losses, all the power generators in current use are AC generators. This means that generating AC power efficiently is a major concern. This section introduces the operation of AC generators and then presents different types of generators used in industry.

A.3.1 Operation of an AC Generator

Electrical engineers across the world have spent over a century improving the design of AC generators, but the basic principle is fairly simple. AC generators are similar to the DC generators discussed earlier, but with one major difference: the output of an AC generator doesn't need to be rectified or smoothed. This means that no commutator is needed to reverse the current's direction. Figure A.6 depicts an example of the output of a single-phase AC generator.

Figure A.6
Unrectified power from
an AC generator

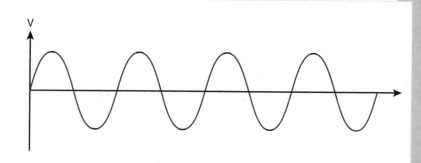

Three characteristics of an AC generator's power can be determined as follows:

- The amplitude of the generated power is proportional to the strength of the field magnets.

- The number of phases in the output power equals the number of conductors rotating between the magnets.

- The frequency of the generated power is determined by the rotational speed of the rotor and the number of poles.

This last point is important to understand. Denoting the rotor's speed as n_s and the number of poles as p, the equation for the frequency of the power produced by an AC generator is given as follows:

$$f = \frac{n_s p}{120}$$

This is essentially the same equation as the one presented in Chapter 6, which computes the speed of a synchronous AC motor. As an example, if the rotor of a four-pole generator turns at 90 rotations per minute, the generated power will have a frequency equal to (90)(4)/120 = 360/120 = 3 Hz.

A.3.2 Magnetos and Self-Excited Generators

There are many ways to classify AC generators, just as there are many ways to classify AC motors. They can be categorized by size, weight, torque, and speed. Some are single-phase whereas others are polyphase.

This section categorizes generators according to the element that generates the magnetic field. If a generator uses permanent magnets to produce its field, it's referred to as a magneto. If it relies on energized coils in the rotor, it's called a self-excited generator.

 note

Some sources use the term *alternator* to refer to self-excited AC generators, whereas other sources employ the term for all AC generators. Because of the confusion, this appendix avoids the term entirely.

Magnetos

In Figure A.1, the bars labeled "N" and "S" represent the poles of a permanent magnet. AC generators with permanent magnets are called magnetos, and they're simple and reliable. For this reason, they're commonly used in applications requiring high reliability, such as in aircraft and lighthouses.

In most magnetos, the magnets are located outside the wires that will receive the induced current. The magnets rotate around the wires, which stay in position. In other words, the magnets are on the rotor and the armature is on the stator. This structure is similar to that of the outrunner brushless DC motor discussed in Chapter 3. This is shown in Figure A.7.

Magnetos are frequently used to provide the current needed by small ignition engines, such as those in lawnmowers and old-fashioned automobiles. In this case, the generator's purpose is to provide isolated pulses of high voltage instead of continuous power.

Self-Excited Generators

Unlike magnetos, self-excited generators create magnetic fields by sending DC current through windings in the rotor called field coils. Put another way, these generators require electrical and mechanical power as input and produce electrical power as output. This process of using electricity to produce electricity is called self-excitation.

Figure A.7
Magneto rotor and stator

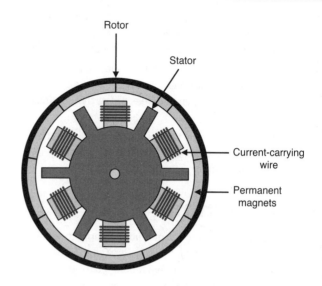

This may seem inefficient, but these generators have one major advantage over magnetos: They can produce a great deal more electrical power. For example, one major application of self-excited generators involves providing power to the electronic components in modern cars and trucks, such as the headlights. These components can collectively require as much as 100 A, which is beyond the capability of a magneto. But with enough battery current and windings in the rotor, a self-excited generator can produce this level of current and more.

To show how this works, Figure A.8 presents the basic structure of a self-excited generator.

Figure A.8
Rotor and stator of a self-excited
generator

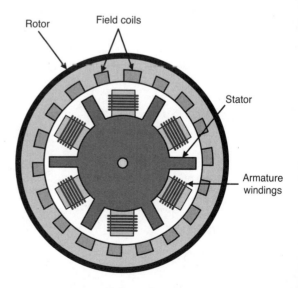

One important use of self-excited generators is within automobiles. In this case, the current to the field windings is provided by the battery. The current produced by the generator is rectified, smoothed, and delivered to the vehicle's electronics. In addition to powering systems such as the headlights and power steering, this generated current also charges the battery.

In addition to powering automobiles, self-excited generators are commonly used to provide electricity to buildings and residences. At the time of this writing, the top five largest power production facilities rely on hydroelectric power, in which water rotates a turbine inside a self-excited generator.

As an example, the largest facility in the world is the Three Gorges Dam, which contains 32 self-excited generators. Each is capable of generating 700 MW of power, and the outer diameter of each generator's stator is 70 feet.

A.4 Summary

Electric motors are fun and interesting, but without electric generators, society as we know it would collapse. Thankfully, if you understand the basics of how motors work, it's easy to understand how generators work. In a motor, a current-carrying conductor in the presence of a magnetic field receives motion. In a generator, a conductor moving in the presence of a magnetic field receives current.

Every generator produces changing levels of current, but DC generators require special circuitry to make the output power suitable for DC devices. The commutator provides rectification, which ensures that all the current flows in one direction. In addition, smoothing circuits are employed to remove variation from the output power.

The majority of the electric generators we encounter in daily life are AC generators. Some AC generators, called magnetos, contain permanent magnets. Self-excited generators have windings in the rotor and stator. These generators are capable of generating much more power than magnetos, and for this reason, they're used in automobiles and large-scale power generating facilities.

GLOSSARY

When you're learning about electric motors, one of the main difficulties involves keeping track of all the terminology. The goal of this glossary is to provide simple definitions for many of the terms used throughout this book.

absolute encoder—An encoder that provides the controller with the motor's speed and position.

AC motor—A motor that runs on AC (alternating current) power. May be synchronous or asynchronous, single-phase or polyphase.

air gap—Space separating the rotor and stator in an electric motor.

aircore motor—Motor whose electromagnets lack iron cores.

Arduino—Family of microcontroller-based circuit boards that are wildly popular among makers because of their simplicity, low cost, and free design.

armature—Conductive element of the motor that carries current.

asynchronous motor—A motor that operates at a different speed than the frequency of the incoming power (frequently a synonym for induction motor).

auto-cutoff—Voltage level at which the ESC will automatically reduce power to the motors.

back-EMF—Voltage generated by an electric motor during its operation. The voltage is proportional to the motor's speed and is oriented to oppose the current entering the motor.

battery eliminator circuit (BEC)—A feature of many ESCs that provides power to the RC receiver. This makes it unnecessary to create a separate connection for the receiver.

bipolar stepper—Stepper with a four-wire connection, where each wire connects to one pole of the internal electromagnets. This stepper requires an H bridge (or a similar circuit) to provide control.

brush—A metal contact, usually between a moving element (such as a rotor) and a stationary element.

brushed motor—A DC motor that uses a commutator to switch the direction of current in the armature. These motors are inexpensive and simple to control, but require maintenance because of the brush.

brushless DC motor (BLDC)—A DC motor controlled with timed pulses of current. The lack of a commutator gives BLDCs high reliability and excellent performance, but they're difficult to control.

closed-loop control system—Control system that provides feedback to the controller.

cogging—The tendency of the rotor to briefly lock in place as it rotates.

commutation—The process of reversing the flow of current with every half-revolution of the rotor. Used in brushed DC motors and DC generators.

copper loss—Power loss caused by the armature's electrical resistance. If the current is I and the armature resistance is R_a, the copper loss equals $I^2 R_a$.

coreless motor—Motor whose electromagnets lack iron cores.

dead bandwidth—Maximum pulse length (in seconds) that a servomotor will ignore. If a pulse's length is greater than the dead bandwidth, the servomotor will rotate in response.

duty cycle—In a train of PWM pulses, duty cycle is the ratio of the width of a pulse to the spacing between the pulses. This value is commonly expressed as a percentage.

dynamometer—Instrument for measuring torque and power, commonly used to measure an electric motor's operational characteristics.

electric speed control (ESC)—Circuit that accepts signals from a controller and delivers power to a DC motor (frequently used with brushless DC motors).

electromagnet—Magnet formed by flowing current through a coil of wire. In a high-strength electromagnet, the wire is wound around an iron core.

encoder—Mechanism that attaches to a motor to provide feedback to a controller. Encoders are usually optical or magnetic.

field coil/field winding—Windings whose purpose is to generate the motor's magnetic field.

field magnet—Permanent magnet(s) whose purpose is to generate the motor's magnetic field.

flyback diode—A diode placed in parallel with a motor to provide a path for current produced by the motor's back-EMF.

fractional slot motor—A motor whose number of windings (slots) is a not a multiple of the number of poles.

H bridge—A circuit with four switches capable of delivering current in two directions. This makes it possible to operate a motor in forward and reverse.

half-step—Method of controlling a stepper motor in which the controller alternates between energizing one winding and two windings. This rotates the stepper half its usual angle, but the torque changes depending on how many windings are energized.

Hall effect sensor—A sensor that changes its output voltage when a magnetic field is present.

hobbyist servo—A DC motor with three connections: power, ground, and control. The PWM pulses control the rotor's angle, and the motor doesn't provide feedback to the controller.

holding torque—Torque exerted by a stepper motor to maintain its angular orientation.

horsepower (hp)—Measurement of power exerted by a motor or engine, where 1 horsepower (hp) equals 745.699872 watts.

hybrid (HY) stepper—Stepper motor that combines characteristics of permanent magnet steppers and variable reluctance steppers to provide high angular resolution and significant torque.

incremental encoder—An encoder that tells the controller the motor's speed but not its position.

inrunner—A brushless DC motor (BLDC) whose rotor is on the inside and whose stator is on the outside. In general, inrunners provide greater speed than outrunners but less torque.

insulated-gate bipolar transistor (IGBT)—A transistor whose resistance between the source and drain terminals changes when voltage is applied to the gate terminal. This is commonly used to switch power to a motor on and off. It doesn't switch as quickly as a MOSFET, but can handle higher current and provides a lower voltage drop when switched on.

integral slot motor—A motor whose number of windings (slots) is a multiple of the number of poles.

inverter or power inverter—A circuit that produces precise waveforms needed to power motors. For BLDCs, inverters generate timed pulses at different voltage levels. For AC motors, inverters generate sinusoids in one or more phases.

Jedlik, Anyos—Hungarian engineer who constructed the first practical electric motor.

Laplace transform—Mathematical transform that converts differential equations (hard) into algebraic equations (easy), and back again.

linear motor—An electric motor that produces linear motion as output, as opposed to a rotary motor, which produces rotation.

lithium-iron-phosphate (LiFePO$_4$ or LFP)—New battery technology used in some motor control applications. LiFePO$_4$ batteries don't provide as much energy as Li-Po batteries, but they're more stable.

lithium-polymer (Li-Po)—Battery technology commonly used in motor control. Li-Po batteries provide excellent energy per weight, but can be unstable if mishandled.

metal-oxide-semiconductor field-effect transistor (MOSFET)—A transistor whose resistance between the source and drain terminals changes when voltage is applied to the gate terminal. This is commonly used to switch power to a motor on and off. It provides faster switching than an IGBT, but presents a greater voltage drop when switched on.

microstep—Method of controlling a stepper motor in which control pulses are split into multiple short pulses, frequently in a sinusoidal pattern.

no-load speed—A DC motor's speed when no load is attached. This is generally the motor's maximum speed. Denoted as ω_n.

Oersted, Hans Christian—Danish physicist who studied the motion of compass needles in the presence of a current-carrying conductor.

open-loop control system—Control system that doesn't provide feedback to the controller.

outrunner—A brushless DC motor (BLDC) whose stator is on the inside and whose rotor is on the outside. In general, outrunners provide greater torque than inrunners but less speed.

peak efficiency point—Operation conditions at which a motor operates with maximum efficiency.

permanent magnet DC motor (PMDC)—DC motor that uses permanent magnets to produce the magnetic field.

phase angle—In electronics, this identifies the angle between AC voltage and current. This is positive if the voltage precedes the current and negative if the current precedes the voltage.

PID (proportional-integral-differential) controller—Controller that computes control signals with a weighted sum of the error's proportional, integral, and differential aspects.

polarity—The relative locations of a magnet's north and south poles.

polyphase motor—An AC motor that receives multiple (usually three) phases of sinusoidal power.

power—Rate at which work is performed. Electrical power equals current times voltage. For rotation, power equals torque times rotational speed. For linear motion, power equals force times velocity.

power factor—Number between 0 and 1 that identifies how much of the input power is being converted to real work. This is important when designing systems with AC motors.

power inverter—See *inverter*.

pulse width modulation (PWM)—Method of providing power to a motor using a train of evenly spaced pulses.

rotary motor—An electric motor that produces rotation as output, as opposed to a linear motor, which produces linear motion.

rotor—The portion of the motor that moves as the motor operates.

RPM—Revolutions per minute, a common measurement for rotational speed, where 1 RPM equals 6°/sec.

sensored motor control—Controlling a motor (generally a BLDC) by reading position data from its sensors (generally a Hall effect sensor).

sensorless motor control—Controlling a motor (generally a BLDC) without any sensors present on the motor. This usually involves monitoring the back-EMF to determine when zero-crossings occur.

series-wound DC motor (SWDC)—A DC motor that uses field windings in series with the armature to produce the magnetic field.

servomotor—A motor that provides feedback to the controller to enable high-precision operation. A servo that doesn't provide feedback is a *hobbyist servo*.

shunt-wound DC motor (SHWDC)—A DC motor that uses field windings in parallel with the armature to produce the magnetic field.

stall torque—A DC motor's torque when the load is too large to turn. This is the maximum torque the motor is capable of exerting. Denoted as τ_s.

stator—The portion of the motor that remains stationary as the motor operates.

step angle—Angle that a stepper motor rotates through with each excitation. Common angles are 30°, 15°, 7.5°, 5°, 2.5°, and 1.8°.

stepper motor—An electric motor (usually brushless DC) that rotates through a specific angle (step angle) and stops.

synchronous motor—A motor that operates at the same speed as the frequency of the incoming power.

torque—Physical quantity similar in concept with rotational force. If the force acting along a circular arc is perpendicular to the circle's radius, the torque equals the product of the force and radius.

torque-speed curve—Curve that identifies a motor's torque at different levels of speed.

unipolar stepper—Stepper with a six-wire connection, where the wires connect to poles or center taps. This is simpler to control than a bipolar stepper, but provides less torque because only half of an electromagnet is being used at any time.

variable reluctance (VR) stepper—Stepper motor whose rotor is a toothed iron disk. It provides high angular resolution but low torque.

watt (W)—Measurement of power, where 1 watt (W) equals 0.00134102209 horsepower (hp).

INDEX

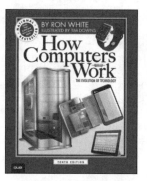
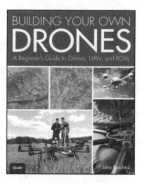
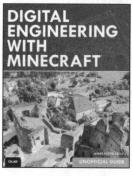

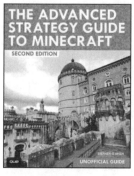